Proteins: The Essential Guide for the Curious Mind.

Proteins: The Essential Guide for the Curious Mind

Bernardo Pérez de Buerres Ramírez

Novo Codex

2026

 During editing, grammar was corrected using AI tools.

Cover photo: Human prion protein, molecular model. Prions are abnormal proteins that cause a group of fatal neurodegenerative diseases, including Bovine Spongiform Encephalopathy (BSE) in cows and Creutzfeldt–Jakob disease (CJD) in humans. Prions lack nucleic acid (RNA or DNA) genomes for replication. Source: Alamy. Under license.

First Edition: July 15th, 2026

ISBN: 979-8-9931872-2-8

Edited by *Novo Codex Editions,* 6 Liberty Square, Suite 2405, Boston, Massachusetts 02109, USA.

This book has been set in Minion Pro, 11-point. Minion Pro is a serif typeface designed by Robert Slimbach in 1990.

www.perezdebuerres.com

Novo Codex

Impreso en EE. UU. — Printed in the USA

A protein biochemist is a psychologist of molecules, observing their behavior in solution and predicting the conditions that transform instability into therapeutic potential.

Contents

Introduction

This book is dedicated to the tireless curiosity that drives us to understand the complex workings of the universe, from the vast cosmic structures to the tiniest molecular machines. It is for the aspiring scientist, the experienced researcher expanding beyond their specialty, and everyone who has ever looked at a living organism and wondered, How does it do that? Most of all, it is for those who believe that life's deepest mysteries are not beyond our reach but waiting to be uncovered through clear explanations, thoughtful questions, and engaging stories. Biochemistry is often seen as a confusing maze: a landscape full of unfamiliar names, abstract diagrams, and seemingly impenetrable ideas. Yet at its core, it tells a story of incredible beauty and importance—the story of proteins. These remarkable molecules are the workhorses of life, the builders, engineers, messengers, and machines that make biology possible. Every breath we take, each step we walk, and every thought we have depend on their tireless work. For too long, proteins have been confined to specialized textbooks, their wonder hidden behind jargon. This book aims to remove that barrier. My goal is to make biochemistry understandable through simple language, vivid analogies, and compelling examples. Abstract ideas will be connected to everyday experiences, like amino acids as beads on a string, protein folding as carefully crafted sculptures, and molecular machines as intricate clockwork operating far beyond our sight. These comparisons do not oversimplify but serve as bridges to make complex concepts easier to understand. A glossary is included to help with comprehension.

My journey as a biochemist has been driven not only by discovery but also by a deep desire to share the awe I feel when I see how life truly works at the molecular level. I have seen how understanding these microscopic marvels can change how we view our own bodies and the living world around us. Much of my professional life has involved teaching protein structure and function to specialists and students willing to wrestle with complexity, equations, and dense terminology. I have used knowledge of protein behavior to develop therapeutic agents. But this book is different. It is written for the curious mind rather than the credentialed one. No previous expertise is needed, only a willingness to wonder. I hope this exploration of proteins will inspire a deeper appreciation for the elegant chemistry behind all existence and encourage a lifelong pursuit of knowledge. Imagine, for a moment, a microscopic world teeming with activity, a bustling metropolis where every citizen has a specific and vital role. This is the world inside our cells. In this city, proteins are everywhere. They are the builders who give cells shape and strength, the couriers who carry messages, the security forces who defend against invaders, and the engines that power movement. They assemble, disassemble, communicate, adapt, and respond, all with astonishing precision. Without proteins, life as we know it would simply cease to exist.

Together, we will explore the fundamental building blocks of proteins: amino acids. We will follow their evolution from simple chains to exquisitely folded three-dimensional structures and examine why shape is essential for function. Along the way, we will examine one of biology's most fascinating phenomena: protein folding. Like intricate origami, proteins must fold correctly to perform their roles. When this molecular choreography succeeds, life flourishes. When it fails, the consequences can be profound, contributing to conditions ranging from neurodegenerative diseases to cancer. By understanding these processes, we gain insight not only into biology but also into health, disease, and the future of medicine. As our understanding of proteins deepens, so too does our ability to harness them. Advances in biotechnology, medicine, and synthetic biology are built on insights into protein structure and function. From life-saving drugs to sustainable materials and innovative therapies, proteins are at the forefront of scientific innovation, and they will play an increasingly central role in shaping our future.

This book is an invitation: to look more closely, to ask questions, and to marvel at the protean performers of life beyond the simplistic view of proteins as mere dietary supplements, revealing them as the true architects and engineers of all living things. Proteins are not just biological components; they are dynamic, responsive, and essential players orchestrating the grand symphony of living systems. By understanding them, we gain not only scientific knowledge but also a deeper appreciation of the delicate, powerful chemistry that unites all living things. Let us begin this journey into the hidden world within us—the world where life's most important work is carried out silently and beautifully, molecule by molecule.

July 9th , 2026
Brookline MA

CHAPTER 1.
Proteins in Everyday Life

We often hear about proteins in the context of diet, especially muscle building. Phrases such as "protein shake," "protein diet," and "high-protein meal" are commonplace, suggesting that proteins serve mainly as structural building materials, like bricks in a wall, or as readily available energy sources. Although proteins do contribute to our physical structure and can be broken down for energy, this view greatly understates their importance. Seeing proteins only as dietary components is like looking at a smartphone and noticing only its casing while overlooking the intricate circuitry and processing power within. In reality, proteins are not passive building blocks; they are dynamic, versatile, and indispensable molecular machines that orchestrate nearly every function in a living cell. Consider the common misconception that protein is solely for muscle building. Muscle tissue is rich in structural proteins such as actin and myosin, but these represent only one aspect of protein function. Actin and myosin are also crucial for cell shape, cell division, and the internal transport of materials in non-muscle cells. Furthermore, dietary protein does not directly "build muscle" in isolation. It provides amino acids, the fundamental building blocks that cellular machinery assembles into the specific proteins required for muscle growth, repair, and countless other functions. The body uses these building blocks with remarkable precision. It does not simply pile them up; it sequences and folds them into intricate three-dimensional structures, each tailored to a particular task. Another misconception is that amino acids from plant proteins are inherently better than those from animal proteins. In fact, amino acids are the same whether they come from plant or animal sources, although some essential amino acids are less abundant in certain plant proteins, as we will review later.

Imagine a bustling city where architects design buildings, construction crews raise them, delivery trucks transport goods, communication networks relay information, and power plants generate energy. Now translate this analogy to the microscopic world of a cell. Proteins perform all these essential roles at once. They are enzymes that catalyze the thousands of chemical reactions required for metabolism. They provide structural components that give cells shape and integrity, forming the scaffolding of life. They act as signaling molecules that allow cells to communicate with one another and with their environment, coordinating complex biological processes. They are molecular motors that drive movement, from the beating of our hearts and the twitching of our muscles to the transport of molecules within cells. They also serve as defenders that protect us from invaders and as machinery that helps replicate our genetic material. The diversity of tasks performed by proteins is staggering, and their importance is fundamental. Within the microscopic confines of a cell, these molecular machines are constantly in motion; they are active and alive. They do not remain static. Instead, they interact, transform, and execute their programmed tasks with remarkable speed and efficiency. Understanding proteins means understanding the engines that drive life itself. A more detailed treatment of proteins as molecular machines is provided in *The Engines of Life: Structure and Function of Proteins at Work* (see Further Reading).

The Essential Amino Acid Challenge in Vegetarian Diets

Proteins are biological macromolecules and essential building blocks of life. At the heart of every protein are amino acids, the small molecular units that determine what a protein can do and how well the body can use it. Of the 20 amino acids that make up the proteins in our bodies, 9 are classified as essential: histidine, isoleucine, leucine, lysine, methionine, phenylalanine, threonine, tryptophan, and valine. They are called "essential" because the body cannot manufacture them on its own. Every day, we must obtain all 9 from food. If even one is consistently missing, the body's ability to build and repair tissues, produce enzymes, and support immune function begins to suffer. This is where the concept of complete proteins becomes important. A complete protein contains all 9 essential amino acids in adequate amounts. Animal-based foods such as meat, fish, eggs, and dairy are naturally complete proteins. Most plant-based foods, however, are incomplete proteins, meaning they are missing or low in at least one essential amino acid. This does not make plant foods unhealthy; far from it. It simply means that people following a strictly plant-based diet need to be thoughtful about combining protein sources to cover all their amino acid needs. Legumes, such as beans, lentils, and peas, provide substantial protein, fiber, and other vital nutrients. Tofu and tempeh, both derived from soybeans, show how plant proteins can be transformed into versatile foods. Grains such as quinoa and oats, along with nuts and seeds, also contribute to daily protein intake. Although many plant proteins are not complete on their own, a varied diet can provide all the necessary amino acids. Diversifying protein sources is not merely a matter of preference; it reflects both the varied protein architectures found in nature and the body's ability to use them.

The pattern of amino acid limitations in plant foods is predictable, and understanding it is the first step toward solving the puzzle. Grains such as rice, wheat, and corn are typically low in lysine but contain adequate amounts of methionine. Legumes, including beans, lentils, chickpeas, and peas, show almost the opposite pattern: they are rich in lysine but lower in methionine. Vegetables, while packed with vitamins, minerals, and fiber, generally do not provide enough essential amino acids on their own to meet daily protein needs. Each plant food group, in other words, has its own amino acid "fingerprint," with strengths in some areas and gaps in others.

The good news is that nature offers an elegant solution: complementary proteins. By combining different plant foods, we can fill one another's amino acid gaps and together obtain all 9 essential amino acids. The classic example is rice and beans, a pairing found in cuisines across the world, from Central and South America to the Middle East. Nutritionally, it is brilliant. Other effective combinations include hummus with pita bread, lentil soup with whole-grain bread, and peanut butter on whole-wheat toast. These pairings have nourished populations for centuries, long before nutritional science could explain why they worked so well.

The reassuring takeaway for anyone following or considering a plant-based diet is this: we do not need to achieve perfect amino acid balance at every meal. Research has shown that, as long as we eat a varied diet of plant proteins throughout the day, the body can pool amino acids from different sources to meet its needs. The key is variety, regularly incorporating grains, legumes, nuts, seeds, and vegetables across meals. A well-planned plant-based diet can provide all 9 essential amino acids and offer many other health benefits. The challenge is real, but with a little knowledge, it is entirely manageable.

Why Eating Collagen Is Not the Same as Building Skin Collagen

Marketing often suggests that ingesting collagen delivers it directly to the skin, producing smoother, younger-looking skin. This idea overlooks how digestion works. Collagen, like all dietary proteins, does not enter the bloodstream intact. It is broken down into amino acids and short peptides. Once digested, the body no longer recognizes it as "collagen for the skin" but as a source of basic building blocks. These amino acids are absorbed and distributed throughout the body, where they are used according to immediate physiological needs. They may be incorporated into muscle proteins, enzymes, hormones, immune molecules, or tissues undergoing repair. The body prioritizes survival and function over cosmetic outcomes, so skin renewal is only one possible destination for these nutrients. Skin aging itself is a complex biological process influenced by genetics, sun exposure, hormonal changes, inflammation, and overall nutrition. Although adequate protein intake is essential for healthy tissues, no single protein source can selectively target the skin. Simply eating collagen does not instruct the body to produce more skin collagen, nor does it reverse the natural decline in collagen production that occurs with age. In this context, collagen supplements are best understood as another form of dietary protein rather than a specialized anti-aging treatment. A balanced diet that provides sufficient protein, vitamins, minerals, and antioxidants supports overall health, including skin health. Thus, the claim that consuming collagen alone can rejuvenate the skin or make a person younger is not supported by basic principles of human physiology.

Gelatin and Jell-O

When you make Jell-O or use gelatin in cooking, you are working with collagen, a structural protein extracted from animal bones, skin, and connective tissue. When gelatin powder dissolves in hot water, the protein chains separate and float freely. As the mixture cools, these long collagen-derived molecules form a loose, three-dimensional network that traps water, creating the familiar wobbly, semi-solid texture. This is why Jell-O melts in your mouth: body heat is enough to break the weak bonds that hold the gel together, releasing the trapped water and returning the material to a liquid state.

Why Eggs Turn White When Cooked: A Light-Scattering Explanation

One of the most obvious examples of protein behavior occurs when we cook an egg. In its raw state, egg white, or albumen, is nearly transparent because its proteins, mainly ovalbumin, exist as individual, tightly folded globular molecules dissolved evenly in water. These native proteins are very small, about a few nanometers, and are spread throughout the liquid. Because they are much smaller than the wavelengths of visible light (400–700 nm), they do not significantly scatter light. Light passes through them much like it does through clear glass, making raw egg white appear transparent or just slightly cloudy. When heated, however, albumen undergoes a structural change. Heat provides enough energy to break the weak bonds—including hydrogen bonds, hydrophobic interactions, and disulfide bridges—that keep each protein in its compact, folded shape, as we will discuss later. The proteins denature: they unfold and reveal regions that were previously hidden. These unfolded proteins become "sticky" because reactive groups are now exposed on their surfaces. They start to bond with nearby unfolded proteins, forming an interconnected three-dimensional network. This aggregation process is irreversible, which is why an egg cannot be "uncooked." The white's appearance results from the size of these protein aggregates relative to the wavelength of light. The tangled protein network creates clumps and irregularities ranging from hundreds of nanometers to several micrometers, sizes comparable to or larger than visible-light wavelengths. When light hits particles of this size, it undergoes Mie scattering, a type of scattering that affects all wavelengths of visible light roughly equally. Unlike Rayleigh scattering, which favors shorter blue wavelengths and explains the blue sky, Mie scattering from larger aggregates scatters red, green, and blue light in roughly equal amounts. When all wavelengths of visible light scatter back toward the eye simultaneously, the brain perceives the result as white. The cooked egg white also contains countless tiny pockets, interfaces, and density variations within the protein matrix. Each boundary marks a change in refractive index—the speed at which light travels through different materials. Light bends and scatters at each interface. In effect, the cooked egg becomes a dense, disordered collection of microscopic scattering centers. The same principle makes snow white, milk white, and clouds white. In each case, the material is not inherently white; repeated scattering of light from countless small structures creates the opaque white appearance.

From Clear Liquid to White Foam: A Protein Story in a Bowl

Have you ever wondered why egg whites—the clear, slightly viscous liquid released from a shell- transform into a brilliant white, cloud-like foam when whisked? It seems almost magical: after a few minutes of vigorous beating, something entirely new appears, with a different color, texture, and behavior. Yet there is no magic here, only elegant chemistry. This everyday kitchen phenomenon is another demonstration of how proteins work, and understanding it opens a window into the molecular processes that shape our food, our bodies, and our lives.

At the core of this transformation is ovalbumin, a protein that makes up about 54% of the protein content in egg whites. In their natural state, ovalbumin molecules are folded into tight, compact shapes, rather like tiny balls of yarn, held together by internal chemical bonds. When we beat egg whites, we introduce mechanical energy into the liquid. This energy can disrupt those internal bonds and cause the proteins to unfold, a process known as denaturation. Think of it as unraveling balls of yarn into long, loose strands. Once unfolded, the proteins are no longer tucked away in compact forms. They become exposed, reactive, and ready to interact with their surroundings in new ways.

As the whisk moves through the liquid, it not only unfolds proteins but also traps air. With each stroke, thousands of tiny air bubbles enter the mixture. The unfolded proteins then rush to the surface of each bubble and form a thin protective coating. In effect, the denatured proteins act as molecular scaffolding, stabilizing each bubble and preventing it from merging with neighboring bubbles or popping. The result is a foam: a stable network of air bubbles suspended in liquid and held together by interlocking protein strands. The more you beat, the more air you incorporate, and the more voluminous and structured the foam becomes.

But why does the mixture turn white? After all, both liquid egg white and the air inside the bubbles are transparent. The answer again lies in the physics of light. When light passes through a clear liquid, it travels mostly in a straight line. When it encounters millions of tiny air bubbles, however, light scatters in all directions, as we saw in the example of cooking an egg. No single ray of light passes through the foam without being deflected many times. The combined effect of this scattering makes the foam appear bright white and opaque, even though neither water nor air has color. The same principle explains why clouds, ocean foam, and whipped cream are white: all contain countless interfaces between air and liquid that scatter light. Ultimately, beaten egg white is a protein-stabilized foam, a remarkable structure made mostly of air and held together by a delicate network of unfolded proteins. This is why beaten egg whites are so valuable in cooking. In meringues, the foam is dried into a crisp, airy treat. In soufflés, the bubbles expand in the oven's heat, producing a dramatic rise. In angel food cake, the foam provides lightness and volume that no other ingredient can. Far from being a simple kitchen trick, beating egg whites is a vivid example of protein versatility and a reminder that invisible molecules quietly shape the textures, colors, and structures of the world around us.

The Science of Baking: How Proteins and Sugar Create Magic in the Oven

When you mix flour, sugar, eggs, and butter to bake cookies or a cake, you are not simply following a recipe. You are orchestrating a complex series of chemical reactions that transform simple ingredients into something delicious. At the heart of this transformation are three key players: proteins, found abundantly in eggs and flour; sugars, including the granulated sugar you add and natural sugars in other ingredients; and the heat of the oven. Eggs serve as crucial binding agents, with their proteins providing structure while their fats add richness and moisture. When you crack an egg into batter, you introduce albumin and other proteins that begin in a liquid state and are transformed by heat into the solid structure that gives baked goods their shape and texture.

As cookies or cake batter enter a hot oven, egg proteins begin to denature and coagulate: they unfold from their original structures and then bond together to form new networks. This process typically starts around 140–150°F (60–65°C) and creates the firm, stable structure that prevents a cake from collapsing into a puddle. Simultaneously, sugars help tenderize the protein structure by interfering with gluten formation, retaining moisture, and lowering the temperature at which proteins coagulate. That's why cakes with more sugar tend to be tender and moist. Sugar essentially gets in the way of proteins forming an overly tight network, which would otherwise result in a tough, rubbery texture.

The most fascinating interaction between proteins and sugars occurs through the *Maillard reaction*, named after the

French chemist Louis-Camille Maillard, who first described it in 1912. The reaction begins with covalent interactions between reducing sugars, such as glucose and fructose, and free amino groups in amino acids, peptides, or proteins at temperatures above 285°F (140°C). Unlike caramelization, which involves only sugars, the Maillard reaction is a complex cascade of hundreds of simultaneous chemical reactions between proteins and sugars. These reactions produce new molecules called melanoidins, which give cookies and cake edges their golden-brown color. This is not merely a surface change; it is a chemical transformation that creates depth, complexity, and richness.

The browning we see on baked goods is a visible sign that flavor is being created at the molecular level. The Maillard reaction produces hundreds of flavor compounds. Some create nutty notes, others add buttery or toasted flavors, and still others contribute caramel-like sweetness or subtle savory undertones. This is why a perfectly browned chocolate chip cookie tastes far more complex and satisfying than raw cookie dough, even though both contain the same ingredients. The golden-brown edges of cookies and the domed top of a cake are not just visually appealing; they are flavor-rich regions where the Maillard reaction has worked most intensely.

What about the irresistible aroma that fills the kitchen when cookies or cake are baking? That is the Maillard reaction announcing itself through volatile compounds that evaporate into the air. These aromatic molecules, including furans, pyrazines, and aldehydes, are byproducts of protein-sugar interactions. The smell of baking is powerful and evocative because it combines many volatile compounds at once: buttery notes from diacetyl, toasted grain aromas from pyrazines, sweet caramel notes from furans, and warm, vanilla-like aromas from aldehydes. This is why the smell of fresh-baked cookies can trigger strong memories and associations. The nose detects the chemical symphony of the Maillard reaction, and the brain connects those molecular signatures with comfort, home, and pleasure. The next time you bake, take a moment to appreciate that you are not just cooking; you are conducting a chemical orchestra in which proteins and sugars work together to create something far greater than the sum of their parts.

Proteins in Beer

Beer is much more than just water, hops, and barley. It is a complex biochemical mixture, with proteins playing a vital role in nearly every aspect of a well-crafted beer. These molecules, mainly from malted barley and sometimes wheat or other grains, affect everything from the beer's appearance in the glass to how it feels on the palate. Although they are present in small amounts, usually between 0.3% and 0.5% of the total composition, their influence is significant and diverse.

When it comes to foam, proteins are indispensable. The creamy, persistent head that crowns a well-poured pint is largely produced by proteins such as lipid transfer proteins (LTPs) and protein Z, both derived from barley. These molecules are remarkably stable. They survive the intense heat of brewing and resist the acidity of the final beer. They migrate to the surface of carbon dioxide (CO_2) bubbles and form protective films that prevent the bubbles from collapsing too quickly. Without these proteins, the beer's head would vanish in seconds, along with much of the aroma that foam helps capture and deliver to the nose.

Proteins also influence flavor and mouthfeel in subtle ways. Medium-sized protein fragments, called polypeptides, contribute to the smooth, full-bodied sensation that distinguishes a rich ale from a thin, watery lager. They interact with other beer compounds, especially tannins from hops and grain husks, creating roundness and depth on the tongue. The same interactions, however, can also produce haze, the cloudy appearance seen in unfiltered wheat beers and New England-style IPAs. In these styles, haze is celebrated as a sign of freshness and juicy flavor. In others, brewers remove it through fining, using agents such as isinglass or bentonite that bind to proteins and pull them out of suspension.

A less desirable feature of protein chemistry is chill haze, a temporary cloudiness that appears when beer is refrigerated and disappears as it warms. This occurs when proteins and polyphenols, plant-based antioxidants from hops and malt, form loose bonds at low temperatures and clump into visible particles. Over time, these bonds can become permanent, producing what brewers call permanent haze, an undesirable trait in clear beer styles. To prevent this, many commercial brewers use silica gel or PVPP, a synthetic polymer, to selectively remove proteins most likely to cause haze without stripping away those responsible for foam and body.

Finally, proteins are essential during fermentation, long before beer reaches the glass. Yeast, the microorganism that converts sugars into alcohol, relies on free amino nitrogen (FAN), a mixture of small proteins and amino acids released from malt during mashing, as a critical nutrient source. Without adequate FAN, yeast can become stressed and produce off-flavors such as acetaldehyde, which smells like green apples, or sulfur compounds. In this sense, proteins are not passive bystanders in the finished product. They actively shape beer from its earliest stages, from the mash tun to the pint glass.

Meat Tenderization, Marinades, and Ceviche

When you marinate meat in acidic ingredients such as lemon juice, vinegar, or yogurt, you are using chemistry to alter proteins. Acid causes muscle proteins, primarily myosin and actin, to denature and unwind, helping tenderize the meat and allowing flavors to penetrate more deeply. Enzymatic tenderizers work similarly. Papain from papaya and bromelain from pineapple are protein-digesting enzymes that break apart tough protein fibers, making meat softer and easier to chew. In ceviche, raw fish is "cooked" without heat by citrus juice. Citric acid denatures fish proteins much as heat would, unfolding them and causing them to aggregate. The fish changes from translucent and soft to opaque and firmer, resembling cooked fish. This demonstrates that protein denaturation is not exclusively a thermal process; changes in pH and mechanical stress can also induce structural transformations.

The Invisible Architecture of Bread

There is something almost miraculous about bread. We start with a sticky, formless lump of flour and water, two simple ingredients, and after kneading and waiting, something amazing happens: the dough swells, rises, and in the oven, transforms into a light, airy loaf with a golden crust and a soft, springy interior. For thousands of years, bakers coaxed this change through instinct and tradition without knowing why it worked. Today, we understand the secret: baking bread is protein chemistry in action, and the main protein network is gluten. But what exactly is gluten? Contrary to what you might think, gluten is not a single protein quietly sitting inside a wheat grain. It doesn't even form until water is added and the dough is kneaded. Wheat flour naturally contains two proteins, glutenin and gliadin, which stay separate and inactive in dry flour. Once water and mechanical action are introduced, these proteins find each other and bond, creating the complex protein network we call gluten. Think of it as a tiny web or net, a detailed three-dimensional mesh of interlocked protein strands that spreads throughout the dough and gives it its stretchy, elastic quality. The more you work the dough, the more extensive and organized this web becomes. The gluten network then acts like a balloon. As the dough rests and rises, yeast eats sugars in the flour and releases carbon dioxide gas through fermentation. In a gluten-free dough, this gas would escape, leaving the bread flat and dense. But the gluten network stretches and expands around each gas bubble, trapping it in place. Millions of these trapped bubbles give risen dough its spongy, airy structure. Without this protein framework, there would be no rise, no lightness, and no loaf—just a flat, unleavened disk. Kneading is the baker's way of building and strengthening this network. When you push, fold, and stretch dough, you're not just mixing ingredients; you're lining up gluten strands, encouraging stronger bonds and a more resilient structure. Under-knead the dough, and the gluten network stays weak and patchy, unable to hold gas well; the result is a dense, crumbly loaf.

Over-knead it, and the strands can become too tight and rigid, making bread that is tough and chewy rather than tender. Skilled bakers develop an almost instinctive sense for the right moment, when the dough becomes smooth, flexible, and springs back gently when pressed, a tactile sign that the gluten network has reached its ideal state.

This is why gluten-free baking is so difficult. Removing gluten causes the loss of the structural framework that makes bread possible. Gluten-free bakers need to use substitutes like xanthan gum, psyllium husk, or alternative proteins from rice, chickpea, or tapioca flour to imitate gluten's properties. These substitutes can get close, but replicating gluten's unique mix of elasticity, strength, and gas-trapping ability remains one of the great challenges of modern food science. The next time you tear open a fresh loaf and admire its soft, airy crumb, remember that you are looking at the work of proteins, invisible molecular architects that have built one of humanity's oldest foods for thousands of years. This is also why people with celiac disease must avoid wheat: their immune systems react negatively to gluten proteins, damaging the lining of their intestines.

Leather Tanning

Leather is essentially preserved animal skin. Tanning stabilizes collagen fibers, preventing decomposition and giving the material durability and flexibility. In this way, a perishable biological tissue becomes a robust, versatile material used for shoes, bags, furniture, and countless other products. The suppleness and strength of a leather wallet or a pair of boots come directly from the protein structure beneath them.

Traditional tanning uses tannins from tree bark or chromium salts that cross-link collagen fibers, making them resistant to bacterial decay, heat, and water. Without this protein stabilization, the hide would rot. Cross-linking also makes leather durable, flexible, and long-lasting.

Blood Clotting

When we get a cut, a cascade of protein interactions helps save our lives. Fibrinogen, a soluble protein in the blood, is converted into fibrin by the enzyme thrombin. Fibrin molecules then polymerize into long, sticky threads that weave together to form a mesh over the wound, trapping blood cells and creating a clot. This elegant protein transformation is why small cuts stop bleeding on their own: the body creates a biological bandage.

Rennet in Cheesemaking

Traditional cheesemaking depends on rennet, a mixture of enzymes whose main active ingredient is chymosin. Traditionally extracted from the stomachs of young calves, chymosin specifically targets κ-casein, a key protein in milk. Milk contains casein molecules arranged into tiny spherical structures called casein micelles. These micelles stay suspended because κ-casein forms a protective, water-attracting layer on their surfaces, which prevents them from sticking together through steric and electrostatic stabilization. Chymosin cuts κ-casein at a specific peptide bond, removing the segment responsible for this protective layer. After κ-casein is cleaved, the micelles lose stability and start to clump together in the presence of calcium ions, creating a three-dimensional protein network that traps water, fat, and other milk components. This process produces semi-solid curds, which serve as the base for most cheeses. Because coagulation involves a targeted enzymatic reaction rather than general protein breakdown, rennet-made curds tend to be firmer, more elastic, and better suited for aging compared to curds formed by acid coagulation. As a result, rennet is essential for many cheeses, including cheddar, gouda, mozzarella, and parmesan, while acid-set cheeses like paneer and ricotta develop different textures, moisture levels, and protein structures.

Yogurt and Fermentation

When bacteria ferment milk to produce yogurt, they break down lactose, the milk' s sugar, and generate lactic acid as a byproduct. As lactic acid accumulates, the milk' s pH gradually drops from about 6. 6.6–6. 6.7 to around 4. 4.6, near the isoelectric point of caseins, the main milk proteins. The isoelectric point is the pH at which a protein has no net charge. Milk proteins mainly consist of caseins, especially αs 1-, αs 2-, β-, and κ- caseins, which are arranged into small structures called casein micelles. In fresh milk, these micelles remain dispersed because κ- casein forms a stabilizing layer on their surfaces, creating electrostatic repulsion and steric hindrance that prevent them from sticking together. As pH decreases during fermentation, negative charges on the casein molecules are gradually neutralized, reducing the repulsive forces that keep the micelles apart. Meanwhile, calcium phosphate inside the micelles becomes more soluble and begins to dissolve, further weakening the micellar structure. Unlike rennet coagulation, which depends on enzyme- driven cleavage of κ- casein, acid coagulation happens through slow chemical changes in charge and mineral balance. As micelles lose stability, they cluster together and form a continuous three- dimensional network of interconnected casein proteins. This network traps water, milk fat globules, minerals, and dissolved substances, turning liquid milk into a thick, semi- solid gel known as yogurt. The casein network gives yogurt its smooth texture, viscosity, and ability to hold shape when spooned. Whey proteins, such as β- lactoglobulin and α- lactalbumin, also influence texture because milk is usually heated before fermentation. Heating partially unfolds these whey proteins, allowing them to interact with κ- casein on the micelle surface. This strengthens the final gel and helps retain water, producing firmer, creamier yogurt with less whey separation. In Greek yogurt, protein content is even higher because much of the liquid whey is drained after fermentation. Straining leaves behind a denser network of casein and whey proteins, creating a thicker texture, increased creaminess, and significantly more protein than regular yogurt. Overall, bacterial acid production, casein clustering, and heat-driven whey protein interactions demonstrate how controlled changes in protein structure can dramatically alter a food' s physical properties.

Wool and Felt Making

Wool fibers are made of keratin, the same protein found in hair and nails. Each wool fiber is covered with tiny scales. When wool is washed in hot water with agitation, heat causes the protein fibers to swell and the scales to open. They then catch on neighboring fibers in a process called felting. That is why a wool sweater can shrink dramatically if washed improperly. Felt makers intentionally use this property, applying heat, moisture, and friction to compact wool fibers into a dense, nonwoven fabric.

Silk and Spider Webs

Luxurious silk fabric is made almost entirely from fibroin, a protein produced by silkworms. These insects spin continuous protein fibers to create cocoons, wrapping themselves in a protective layer as they undergo metamorphosis. Humans have learned to harvest these delicate fibers by unraveling the cocoons and weaving the strands into smooth, lustrous fabric prized for thousands of years. The natural sheen and softness of silk come from fibroin's structure, which reflects light and gives the material its characteristic glow.

Similarly, spider silk is made of specialized proteins called spidroins that are extremely strong and adaptable. Pound for pound, spider silk exceeds the strength of steel and is tougher than Kevlar, making it one of the most impressive natural materials known. Its strength comes from densely packed crystalline regions mixed with flexible, elastic segments. This structure allows the silk to absorb energy without breaking. As a result, spider silk can stretch greatly while keeping its integrity, a combination of features that scientists study to develop advanced materials inspired by natural fibers.

Enzyme Cleaners and Laundry Detergents

Many modern laundry detergents contain proteases, enzymes that are themselves proteins and break down protein-based stains such as blood, grass, sweat, and food. These biological catalysts cleave protein molecules into smaller pieces that are more easily washed away in water. That is why enzyme detergents work best in warm, not hot, water: excessive heat destroys protease enzymes, preventing them from functioning. Proteases are also used in contact lens cleaners, where they remove protein buildup from tears on lenses.

CHAPTER 2.

The Architects of Life: An Introduction to Proteins.

Let us examine several notable roles of proteins in greater detail, beginning with digestion. When we eat, a cascade of protein-based enzymes goes to work. Pepsin in the stomach, along with trypsin and chymotrypsin in the small intestine, breaks down dietary proteins into their constituent amino acids. Without these enzymes, much of the protein we consume would pass through the digestive system largely undigested, and we would be unable to absorb the amino acids needed to build and maintain our own bodies. This highlights a critical point: proteins not only form biological structures but also enable the processes that supply the building blocks for those structures.

Among the body's defense systems, antibodies provide a clear example of proteins functioning as intricate security guards. These Y-shaped molecules circulate through the bloodstream and tissues, identifying and neutralizing foreign invaders such as bacteria and viruses. Each antibody is highly specific, recognizing a unique molecular signature on a pathogen much as a key fits a particular lock. Once bound, antibodies mark the invader for destruction by other immune cells or directly neutralize its harmful effects. This protein-based defense illustrates the power of molecular recognition and offers constant, vigilant protection in a world filled with microscopic threats.

The act of movement, fundamental to our experience of life, is also orchestrated by proteins. In muscles, actin and myosin engage in a remarkable molecular dance. Myosin "heads" attach to actin filaments, pull them, release, and reattach, causing muscle fibers to shorten and generate force. This process is powered by adenosine triphosphate (ATP), the cell's main energy currency, which releases usable energy when its high-energy bonds are broken. Actin and myosin enable us to walk, run, jump, and breathe. Their role is not limited to skeletal muscle: related mechanisms operate in the beating heart, in the movement of cilia lining the airways, and in the transport of vesicles within cells. Proteins are, therefore, the tiny motors that drive nearly all biological motion. Proteins also serve as the structural backbone of cells and tissues. Collagen, the most abundant protein in the human body, forms a strong fibrous network that supports skin, tendons, ligaments, and bones. Elastin allows tissues such as skin and blood vessels to stretch and recoil. Inside cells, actin and tubulin form cytoskeletal networks that maintain cell shape, facilitate cell division, and serve as internal highways for molecular transport. Together, these structural proteins provide the framework that enables cells and organisms to withstand physical stress and maintain their form.

Even the transmission of information, the basis of communication within and between cells, depends heavily on proteins. Cell-surface receptors bind signaling molecules and convert those binding events into intracellular responses. Without this intricate network of protein-based signaling, coordinated multicellular life would be impossible, from the regulation of blood sugar and blood pressure to the complex processes underlying thought and memory.

The essence of every protein, the remarkable molecular machinery that sustains life, lies in its basic building blocks: amino acids. These are not generic components; they are diverse molecular units with unique chemical identities. Think of a palette of twenty colors or an alphabet of twenty letters. The precise arrangement of these twenty common amino acids creates the incredible diversity of proteins found in living organisms. Each protein, whether it's an enzyme that speeds up a reaction, a structural element providing support, or a signaling molecule transmitting information, is a carefully ordered sequence of amino acid units.

Each of the twenty amino acids shares a common architectural blueprint. At its core is a central carbon atom, often called the alpha carbon (Figure 1). Attached to this atom are four components: a carboxyl group (-COOH), an amino group (-NH2), a hydrogen atom (-H), and a unique side chain, or R group. This R group distinguishes one amino acid from another and gives each its characteristic chemical properties. Some R groups are hydrophobic, tending to avoid water, whereas others are hydrophilic and interact favorably with water. Some carry positive or negative electric charges, while others are neutral. Polar amino acids have side chains that interact well with water because they carry partial or full charges. This range of chemical properties enables proteins to interact with diverse molecules, fold into specific three-dimensional structures, and perform their many cellular tasks.

The twenty common amino acids can be grouped broadly by their R-group properties. Nonpolar, aliphatic amino acids such as glycine, alanine, valine, leucine, and isoleucine are often found in protein interiors, protected from water. Aromatic amino acids, including phenylalanine, tyrosine, and tryptophan, contain stable ring structures with delocalized electrons that can support stacking interactions and absorb ultraviolet light.

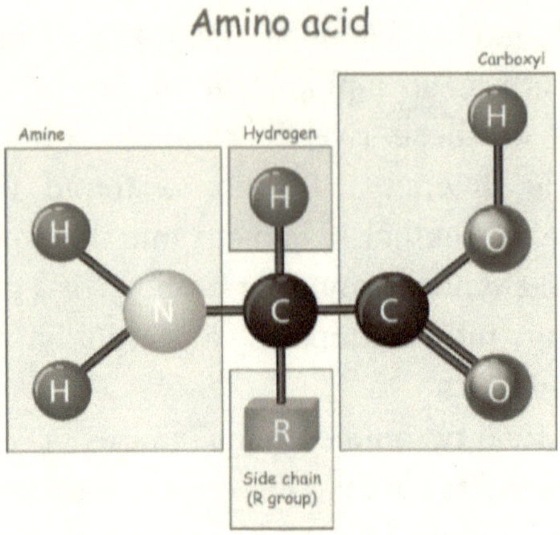

Source: Alamy, under license.

Figure 1. Generic amino acid structure.

Polar amino acids, including serine, threonine, cysteine, asparagine, and glutamine, have R groups that can form hydrogen bonds and are often found on protein surfaces, where they interact with the aqueous cellular environment. Acidic amino acids, such as aspartate and glutamate, carry a negative charge at physiological pH, whereas basic amino acids, such as lysine, arginine, and histidine, carry a positive charge. Cysteine is especially notable because it can form disulfide bonds (-S-S-) with another cysteine residue, creating strong cross-links that stabilize protein structure, as we will see later. Amino acids are linked in a specific order, like beads on a string, to form long chains called polypeptides. The bond connecting one amino acid to the next is the peptide bond, formed when the carboxyl group of one amino acid reacts with the amino group of another, releasing a water molecule (Figure 2). This dehydration reaction is fundamental to biochemistry.

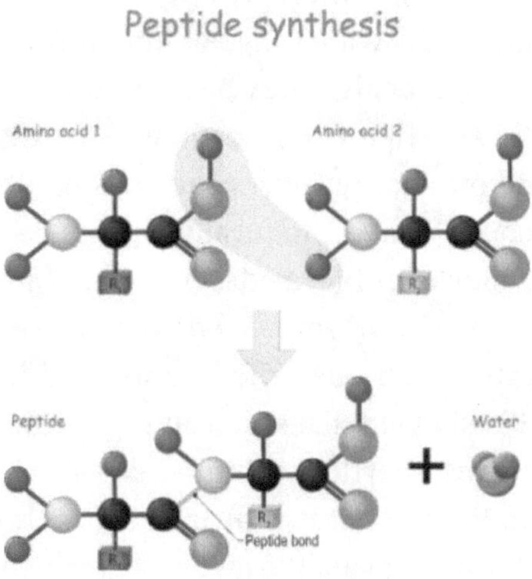

Source: Alamy Stock Vector. Reference: 0Y72287324.

Figure 2. Peptide synthesis. Two amino acids combine to form a peptide, releasing a water molecule and forming a peptide bond.

Once linked, the resulting structure is a *dipeptide*; as more amino acids are added, it forms a *polypeptide chain*.

The sequence of these amino acids in the polypeptide chain is absolutely critical. A change in even a single amino acid can sometimes have profound effects on the protein's structure and activity, leading to disease in some cases.

Think of LEGO bricks as an analogy. A set of twenty unique brick types, each with a different shape and color, can be connected in countless ways to create structures from a simple car to an intricate castle. The order of the bricks determines the final object. Similarly, the amino acid sequence, dictated by the genetic code, determines the final protein. The same twenty amino acids make all proteins in the body, from digestive enzymes and infection-fighting antibodies to actin and myosin in muscle. The ability to arrange these twenty units in specific sequences gives proteins their wide range of functions.

The origin of amino acids dates back to Earth's earliest days. Scientists hypothesize that early Earth, with its distinctive atmosphere and abundant energy sources such as lightning and volcanic activity, could have naturally produced these fundamental building blocks of life, as described in *The Engines of Life: Structure and Function of Proteins at Work* (see Further Reading). Experiments such as the famous Miller-Urey experiment in the 1950s showed that simple inorganic molecules exposed to simulated early-Earth conditions can form various amino acids. These findings suggest that the raw materials for life's complex molecules were available on early Earth, providing a crucial starting point for the emergence of life. These primordial amino acids, formed through abiotic processes, likely accumulated in early oceans, creating a "primordial soup" from which the first self-replicating molecules and, eventually, cells could arise.

The twenty amino acids commonly found in proteins are known as the standard, or proteinogenic, amino acid set. Although other amino acids occur in nature, these twenty are incorporated into proteins by the cellular machinery of protein synthesis under the guidance of the genetic code. Their near universality across life on Earth is powerful evidence of their fundamental importance and of the shared evolutionary history of living organisms. From the simplest bacterium to the most complex human, proteins are written in the same alphabet of twenty amino acids.

The order of amino acids in a protein is not random. It reflects the genetic code, the set of rules cells use to translate genetic information in DNA or RNA into amino acid sequences. The properties of R groups are not limited to their interactions with water; they also determine how a polypeptide chain folds into a functional three-dimensional shape. Hydrophobic R groups tend to cluster inside the protein, away from water, while hydrophilic and charged R groups are more often exposed on the surface, where they interact with water molecules. The balance of attractive and repulsive forces among R groups causes the chain to fold into a specific, stable structure, which is essential for biological activity.

A protein that fails to fold correctly will likely lose its intended function. Thus, the primary structure is the fundamental determinant of subsequent folding. It is like a folded piece of paper: the creases and final shape are determined by the properties of the original sheet and by its manipulation. In proteins, the "manipulation" is driven by the chemical properties of amino acid side chains and by their interactions with one another and with the surrounding solvent. These interactions lead to secondary structures, local folding patterns stabilized by hydrogen bonds between backbone atoms, which we will review in detail later. Secondary structures then assemble into the overall three-dimensional tertiary structure. In some cases, multiple polypeptide chains, each with its own tertiary structure, associate to form a functional quaternary structure.

The importance of the exact amino acid sequence cannot be overstated. It is the blueprint for protein function. Even a tiny change can significantly alter a protein's role. The analogy of amino acids as letters in a biological alphabet helps illustrate their coding potential. With 26 letters, we can create countless words, sentences, and books. With 20 amino acids, the number of possible polypeptide sequences is astronomical. For a protein of just 100 amino acids, the number of possible sequences is 20 to the power of 100, a value so enormous that it is difficult to grasp. Yet nature has efficiently navigated this vast combinatorial space, selecting and refining sequences that perform precise, essential functions. These sequences are not random; they have been shaped by millions of years of evolution to perform specific tasks with remarkable accuracy.

R-group diversity is not defined only by polarity and charge. Some R groups are bulky and can create steric hindrance, influencing how a protein interacts with other molecules. Others are small and flexible, allowing greater freedom of movement within the structure. Histidine is especially interesting because it can act as either an acid or a base near physiological pH, making it crucial for the catalytic activity of many enzymes involved in proton-transfer reactions. An acid donates a hydrogen ion, whereas a base accepts one. Because histidine can act as both a base and an acid, it helps many enzymes move hydrogen ions during chemical reactions. In summary, proteins are built from twenty common amino acids, each with a unique side chain. These amino acids are linked in specific sequences to form polypeptide chains. The precise sequence, or primary structure, is dictated by genes and determines the protein's three-dimensional shape and function. The diversity of side chains provides the chemical variability required for proteins to perform an astonishing range of tasks. From their potential abiotic origins on early Earth to their central role in modern biology, amino acids are the indispensable building blocks of the molecular machines we call proteins.

The journey from a linear amino acid sequence to a functional protein is a remarkable transformation, like a simple thread woven into an intricate tapestry or a string coiled into a complex knot. The primary structure is only the starting material. What gives a protein its specific capabilities is its sophisticated three-dimensional architecture. Each amino acid, through its unique side chain, contributes to a complex interplay of attractions and repulsions that guides the polypeptide chain into a precise, stable, and biologically active conformation.

Imagine holding a long, flexible piece of string. If left alone, it takes on a loose, disordered shape. If twisted, bent, and looped, however, it can become a knot, a coiled spring, or even a three-dimensional sculpture. Proteins undergo a similar process of self-assembly, but instead of being shaped by hands, they are guided by chemical interactions among amino acid side chains and their environment. Hydrophobic collapse is often a key force early in protein folding, as hydrophobic amino acids tend to orient their side chains toward the protein interior, shielding them from water (reviewed in Chapter 4).

This folding process is dynamic. Proteins are not rigid structures; they continuously move and subtly change shape. This flexibility is crucial for their function. For example, an enzyme often bends at its active site, the region where a substrate binds, to fit the substrate, catalyze the reaction, and release the product. Structural proteins like collagen need some flexibility to withstand mechanical stress. Signaling proteins also experience conformational changes when they bind to other molecules, acting as molecular switches that control cellular processes.

A protein's three-dimensional arrangement is often described in terms of hierarchical structures. The primary structure is the linear amino acid sequence. From this sequence emerge secondary structures, localized and recurring folding patterns. The most common are the α-helix, a spiral arrangement stabilized by hydrogen bonds along the polypeptide backbone, and the β-sheet, a more planar structure formed by hydrogen bonds between adjacent strands of the chain (Figure 3). These strands can align in parallel or antiparallel orientations, as discussed in more detail in Chapter 3.

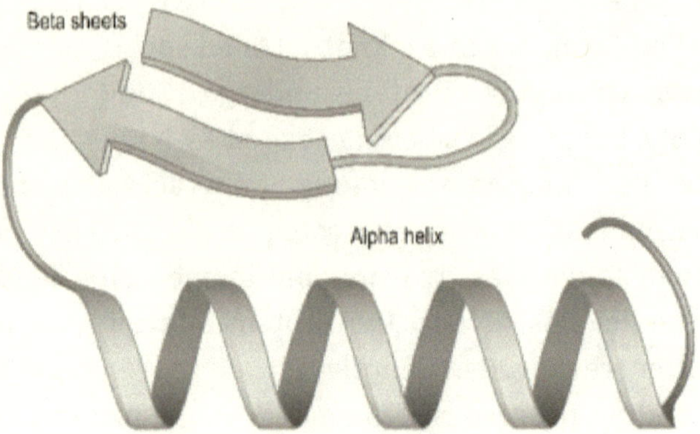

Source: Alamy. Reference OY0111305487

Figure 3. Protein secondary structure

These secondary structural elements come together to form the overall three-dimensional shape of a single polypeptide chain, known as the tertiary structure. This is where the unique shape is created. The α-helices and β-sheets fold and pack together, forming globular domains or elongated fibrous structures. Amino acid R groups, brought close by folding, stabilize this tertiary structure through non-covalent interactions and sometimes covalent disulfide bonds between cysteine residues. Their placement on the protein's surface and inside influences how the protein interacts with substrates, cofactors, other proteins, and other molecules.

For some proteins, the story does not end with a single folded chain. Many functional proteins consist of multiple polypeptide subunits that assemble into a larger functional unit called the quaternary structure. Hemoglobin, the oxygen-carrying protein in blood, is a classic example (Figure 4). It consists of four polypeptide subunits.

Figure 4. Hemoglobin quaternary structure.

The exact arrangement and interactions among these subunits are crucial for hemoglobin's ability to efficiently bind and release oxygen. This quaternary structure enables cooperative oxygen binding: when one oxygen molecule binds to a subunit, it facilitates oxygen binding by the other subunits. This complex interaction at the quaternary level fine-tunes the protein's function. The transition from a linear chain to a functional three-dimensional structure is not passive coiling. It is a thermodynamically driven process that tends toward a low-energy state. In the simplest terms, a protein tends to settle into the shape that is most comfortable and stable, much like a ball rolling downhill until it reaches a low spot where it can rest. The first law of thermodynamics states that energy cannot be created or destroyed; it can only be transferred. When a protein folds, different parts of the molecule form interactions with each other and with the surrounding water. Some interactions release energy, and some require energy. The final folded shape reflects a balance of these energy changes, much like balancing a budget of gains and losses. The second law of thermodynamics says that nature tends to move toward arrangements that are overall more probable. This is often described in terms of entropy, which can be thought of as the number of possible ways things can be arranged. A folded protein is actually more ordered than an unfolded one, but when it folds, many water molecules around it become freer to move. The increase in the water's freedom can outweigh the loss of the protein's freedom, making folding overall favorable. However, the energy landscape of protein folding is complex, with many possible intermediate states. The amino acid sequence determines this landscape and guides the chain through folding pathways, as reviewed in Chapter 4.

The functional importance of precise three-dimensional structure cannot be overstated. A protein's shape determines how it interacts with other molecules. For example, an enzyme's active site is a precisely shaped pocket that complements the substrate. This "lock and key," or more accurately "induced fit," mechanism ensures that the enzyme binds the correct substrate and catalyzes the intended reaction with high specificity. If the enzyme's shape changes, perhaps because of a mutation in its amino acid sequence, its ability to bind the substrate and perform its catalytic function can be severely impaired.

When foreign invaders such as bacteria, viruses, or fungi breach the body's defenses, proteins (especially antibodies) mount a targeted counterattack. Antibodies are specialized proteins produced by immune cells that recognize and bind specific molecular structures on pathogens, called antigens. This binding is highly specific, like a lock-and-key interaction, ensuring that antibodies target foreign entities. The immune system can generate enormous antibody diversity, allowing it to respond to a wide range of threats throughout life. This protein-based defense system is critical for survival.

Proteins are also central to cellular communication. Receptors, typically found on the cell surface or within the cell, bind specific signaling molecules such as hormones or neurotransmitters. Binding causes a shape change in the receptor, initiating intracellular events that may alter gene expression, adjust metabolism, or trigger cell division. This network of protein-driven signaling pathways helps cells coordinate their activities and maintain balance in multicellular organisms. For example, when adrenaline binds to its receptor on heart cells, it increases the heart rate. When insulin binds to its receptor on many cell surfaces, the receptor changes shape and becomes active. It then adds phosphate groups to specific amino acids, mainly tyrosines, in a process called autophosphorylation. These phosphate groups serve as signals that recruit other proteins and initiate a signaling cascade, increasing glucose uptake from the blood and promoting nutrient storage. Other receptors, such as the acetylcholine receptor, function directly as ion channels. When acetylcholine binds, the channel opens and allows ions, mainly sodium (Na^+), to flow into the cell, changing the membrane's electrical charge and, in muscle cells, triggering contraction. Cholesterol, a fatty molecule common in cell membranes, can influence the stability and activity of this receptor. Studies using receptors isolated from the electric organ of *Torpedo californica* showed that cholesterol does more than provide membrane support; it helps fine-tune protein behavior, stability, and activity. This insight shifted cell biology from viewing membranes as simple barriers to recognizing them as dynamic environments where lipids and proteins work together. HDL receptors, especially on liver cells, help remove cholesterol from HDL particles. When HDL binds to receptors such as SR-BI, cholesterol is selectively transferred into the cell, where it can be used to make hormones, build membranes, or be converted into bile acids for removal. Albumin, another abundant plasma protein, binds and transports fatty acids, hormones, and other molecules while helping maintain osmotic pressure.

Furthermore, dynamic movements within protein structures enable allosteric regulation, a process in which the binding of a molecule at one site on a protein influences its activity at a different site. This is a crucial mechanism for controlling enzyme activity and metabolic pathways. For example, binding of an *allosteric inhibitor* can induce a conformational change that reduces the enzyme's active-site affinity for its substrate, thereby slowing the reaction. Conversely, an *allosteric activator* can stabilize a more active conformation of the enzyme. These dynamic shifts in protein shape are central to how biological systems fine-tune their responses.

The study of protein structure and dynamics has been transformed by techniques such as X-ray crystallography, nuclear magnetic resonance (NMR) spectroscopy, and cryo-electron microscopy. These methods allow scientists to visualize proteins at atomic or near-atomic detail, revealing their complex shapes and capturing aspects of their movement. Computational approaches, including molecular dynamics simulations, complement these experimental tools by modeling protein motion over time and clarifying how structure relates to function. Together, these methods show that a linear chain is only the beginning of a protein's journey. The transition from a simple polypeptide to a functional molecular machine illustrates the power of self-assembly and the chemical language at the core of life (Figure 5).

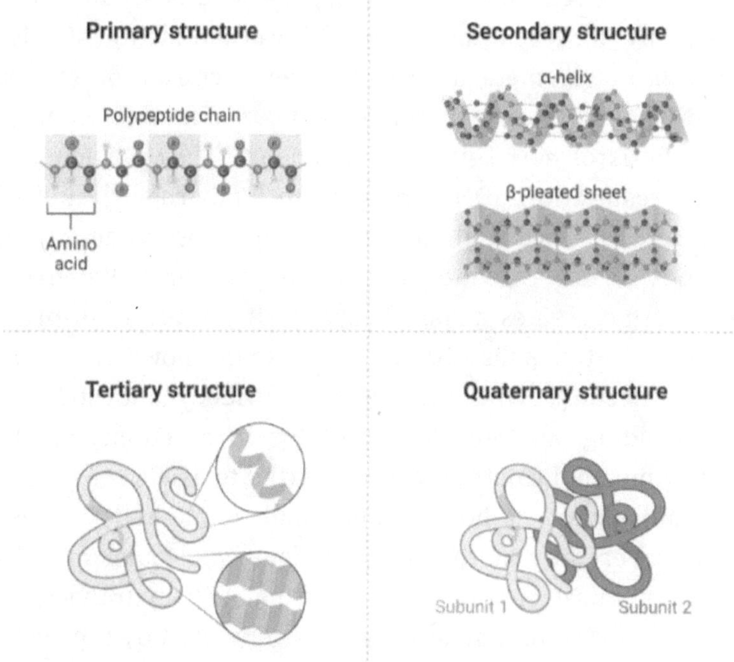

Figure 5. Summary of the types of structures found in a protein (Source: Biorender).

From the microscopic dance of molecules within a single cell to the coordinated actions of entire organisms, proteins are essential agents of life. Their versatility shows the elegance of molecular design, shaped by amino acid sequences and complex three-dimensional structures. We have explored how linear chains fold into functional shapes; now the greater wonder lies in what those shapes do. Proteins are not just building blocks. They are catalysts, messengers, movers, and active players in the symphony of life.

Perhaps the most widely recognized role of proteins is their function as enzymes. Countless biochemical reactions must occur within a cell for it to survive, grow, and reproduce. If left alone, many of these reactions would proceed far too slowly to sustain life. Enzymes are the master chemists of the cell, facilitating everything from the breakdown of food for energy to the synthesis of new cellular components. They participate in cellular respiration, DNA replication and repair, and the production of essential molecules such as vitamins and hormones. Each enzyme has a unique three-dimensional structure that creates a specific active site, a pocket or groove shaped to bind its substrate. This precise fit allows enzymes to catalyze reactions with remarkable efficiency and specificity while avoiding wasteful or harmful side reactions. Without enzymes, metabolic pathways would grind to a halt.

Movement at both cellular and whole-body scales is largely powered by motor proteins. These molecular machines convert chemical energy, typically from ATP hydrolysis, into mechanical force. In a muscle, force is generated by the cyclical interaction of myosin heads with actin. Myosin undergoes conformational changes that allow it to "walk" along actin filaments, producing movements from the subtle twitch of an eyelid to the stride of an athlete. Motor proteins are not limited to muscle. Within cells, kinesin and dynein move along microtubule tracks, transporting organelles, vesicles, and other cargo to their destinations. This intracellular traffic system is vital for cellular organization and function.

Proteins also maintain structural stability and support cellular function. Just as steel beams support a skyscraper, specific proteins provide scaffolding for cells and tissues. Collagen is a striking example. Its triple-helix structure provides exceptional tensile strength, allowing tissues to withstand stretching and mechanical stress. Collagen stands apart through its distinctive architecture: three polypeptide chains wind around one another like a molecular rope, forming one of nature's strongest biological materials. This structure depends on collagen's unusual sequence, in which every third amino acid is glycine, the smallest amino acid, allowing the chains to pack tightly. Collagen is also rich in proline and hydroxyproline, residues that restrict flexibility and stabilize the triple helix. The repeating Gly-X-Y motif, where Gly is glycine, X is often proline, and Y is often hydroxyproline, is collagen's molecular signature.

Beyond its structure, collagen undergoes complex post-translational modifications. After the initial protein chains are synthesized, specific proline and lysine residues are hydroxylated in a vitamin C-dependent process that is essential for stability; this explains why vitamin C deficiency causes scurvy, a disease characterized by defective collagen. Some modified residues are then glycosylated with sugar molecules, adding another layer of reinforcement. Once secreted into the extracellular space, collagen molecules are enzymatically cross-linked to form large, insoluble fibers that withstand enormous mechanical stress. These features make collagen the most abundant protein in the human body, accounting for about 25–35% of total protein and serving as the primary structural scaffold for skin, bones, tendons, ligaments, and blood vessels. Unlike enzymes or signaling proteins, collagen's role is fundamentally architectural: it is the framework that helps hold tissues together and enables movement, protection, and repair.

Another essential structural protein is keratin, found in hair, nails, and the outer layer of the skin, providing durability and protection. Inside cells, proteins like actin and tubulin form complex networks of filaments that offer internal support, preserve cell shape, and are vital for cell division and movement. This structural function is not just passive; these proteins are dynamic, capable of assembling and disassembling as needed, allowing cells and tissues to adapt to and respond to their environment.

In essence, proteins are the multifaceted workhorses of life, involved in nearly every biological process. Later chapters will delve deeper into the specifics of these diverse roles, but this overview serves as a crucial reminder of the central and indispensable nature of proteins in the grand, intricate tapestry of biology.

Proteins are much more than just molecular machines inside cells; they are part of our daily lives, affecting what we eat, how we feel, and the materials around us. When we think of proteins, we might picture protein bars, chicken, or milk. These are direct dietary sources, providing the building blocks the body uses to build tissues, repair damage, produce hormones, and support biochemical reactions.

Animal-based foods such as meat, fish, poultry, eggs, and dairy are known for their high protein content. They often provide complete proteins, meaning they contain all the essential amino acids the human body cannot produce. A serving of beef, for example, contains myoglobin, which stores oxygen in muscle, as well as myosin and actin, the motor proteins involved in contraction. During digestion, enzymes such as pepsin and trypsin break these complex proteins into smaller peptides and amino acids. Cells then use the absorbed amino acids to build the specific proteins needed for muscle growth, enzyme production, immune defense, and many other functions.

Protein's dietary role extends far beyond animal products. Plant-based foods are also important sources, often with distinctive amino acid profiles. The journey from plate to cell highlights the function of proteins throughout the body. Digestion itself is driven by protein enzymes, and proteins also support nutrient absorption, transport, and use. Transport proteins in the intestinal lining move absorbed amino acids into the bloodstream. Once circulating, other proteins, such as albumin in blood plasma, carry fatty acids, hormones, drugs, and other molecules to target sites throughout the body, ensuring that cells receive the nutrients and signals they need.

The structure and movement of our bodies depend fundamentally on proteins. Collagen provides strength and elasticity to skin, forms tough tendons and ligaments, contributes to cartilage in joints, and supports the structural framework of bones. Proteins also influence the materials we use every day. Silk, produced by silkworms and spiders, is renowned for its strength and elasticity. It is composed mainly of fibroin, a protein that forms strong, interlocking β-sheet structures. This arrangement allows silk to absorb shock and resist stretching, making it valuable for applications from clothing to surgical sutures.

Similarly, the properties of wool, a staple in clothing and textiles, are due to keratin, the same protein found in our hair and nails. Keratin proteins are arranged in a complex helical structure that provides wool with its characteristic resilience, warmth, and moisture-wicking properties. When we wear a wool sweater on a cold day, we are benefiting from the unique structural properties of this protein.

The many ways proteins appear in daily life can seem overwhelming, but they underscore their central biological role. Consider healing. When we get a cut or scrape, proteins coordinate repair. Platelets initiate clotting and release protein-based signals. Fibrin forms a protein mesh that closes the wound. Soon afterward, fibroblasts migrate to the injury site and deposit collagen, creating new connective tissue that strengthens the damaged area. Even scar tissue results from how collagen fibers are laid down during healing. This vital repair process is powered by proteins.

Furthermore, the color of our skin, hair, and eyes is largely determined by proteins. Melanin, the pigment responsible for these colors, is synthesized through a series of biochemical reactions catalyzed by enzymes, all of which are proteins. The amount and type of melanin produced, influenced by genetics, dictate the spectrum of human pigmentation. This protein-based coloration not only contributes to our appearance but also may protect the skin from the damaging effects of ultraviolet radiation.

Our sensory experiences also rely heavily on proteins. Vision, hearing, smell, taste, and touch are mediated by specialized receptor proteins. In vision, rhodopsin in rod cells is essential for low-light vision. When light strikes rhodopsin, it triggers a sequence of molecular events that sends a signal to the brain. Similarly, proteins in the ear help convert sound vibrations into electrical signals, taste receptors detect flavors, and receptor proteins in the nasal passages bind odor molecules. These proteins are the gatekeepers of the sensory world, translating external stimuli into signals the brain can interpret.

The act of thinking and processing information in our brains involves proteins. Neurotransmitters, the chemical messengers that enable neurons to communicate, are made and broken down by protein enzymes. Receptors on the surface of neurons are proteins that bind to these neurotransmitters, triggering electrical signals. The structure of neurons, including their complex branching patterns and synapse formation (the junctions between neurons), is supported and maintained by cytoskeletal proteins. The complex cognitive functions we associate with the brain, such as memory, learning, and consciousness, are all supported by these detailed protein-based interactions and structures.

When we consider the materials we use, the importance of proteins becomes even clearer in fields such as biomaterials and medicine. Developing artificial tissues, surgical implants, and drug delivery systems often depends on the biocompatibility and unique properties of proteins. For instance, collagen-based materials are used in wound dressings and tissue-engineering scaffolds because they mimic the natural extracellular matrix of native tissues, thereby promoting cell growth and integration. Protein-based hydrogels are also being studied for targeted drug delivery, releasing therapeutic agents in a controlled way.

In essence, an ordinary day involves an astonishing interplay of proteins. From the food we eat and the air we breathe to the ways our bodies function, communicate, defend themselves, and heal, proteins are fundamental agents of life. Their continuous work connects the microscopic dance of molecules to the visible reality of our daily experience.

CHAPTER 3.

The Blueprint and the Build: Protein Structure Revealed

As noted in the previous chapter, the fundamental blueprint for every protein resides in its primary structure: the linear, unbranched chain of amino acids. This specific sequence—the exact order in which amino acids are linked—is not random. Genes within DNA contain instructions, written in the molecular language of nucleotides, that precisely dictate which amino acid follows the next in the growing polypeptide chain. This is the most basic level of protein information, the starting point from which all later complexities of protein form and function arise. Without the correct sequence, the protein cannot be built properly, and its ability to perform its intended role may be compromised or lost.

The relationship between the genetic code and the amino acid sequence is a central concept in molecular biology, illustrating the flow of information from DNA to RNA to protein (Figure 6). The DNA blueprint is transcribed into messenger RNA (mRNA), which then exits the nucleus and moves to the ribosome, the cellular machinery responsible for protein synthesis. At the ribosome (Figure 7), the mRNA sequence is read in three-nucleotide units called codons. Each codon corresponds to a specific amino acid or to a stop signal.

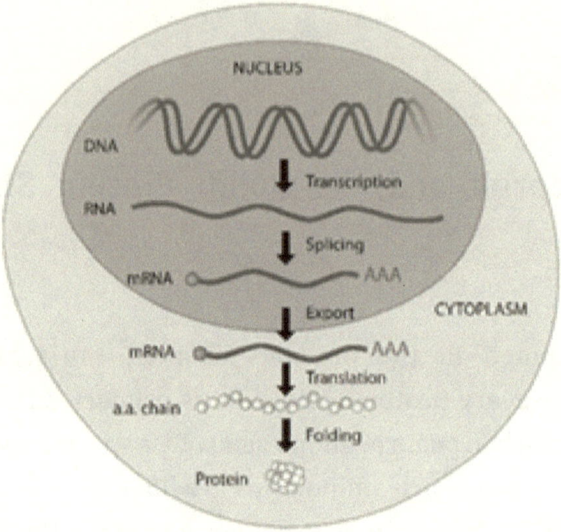

Source: Alamy under license.

Figure 6. Gene expression: From DNA to Proteins.

This remarkable system ensures that the correct amino acid is added to the growing polypeptide chain at each step of protein synthesis. For example, the codon AUG encodes methionine and often serves as the start signal for translation. Other codons, such as UUU, encode phenylalanine, while GGU encodes glycine.

Although GUG usually encodes valine, this detailed mapping, refined through evolution, ensures that genetic information in DNA is accurately translated into a specific amino acid sequence. For example, if the mRNA sequence is AUG-GCC-UUC-UGG-UAA, the ribosome recruits the appropriate aminoacyl-tRNAs, each carrying its assigned amino acid, to assemble the polypeptide chain. AUG signals the start and encodes methionine; GCC encodes alanine; UUC encodes phenylalanine; UGG encodes tryptophan; and UAA is a stop codon that ends translation. The resulting chain is methionine-alanine-phenylalanine-tryptophan. This sequence is the primary structure, the first level of protein organization, and the foundation for all subsequent folding and function.

Polypeptides emerging from the ribosome are then folded into their final conformations. Protein synthesis, or translation, converts the genetic code into an amino acid sequence through a complex and highly regulated cellular process.

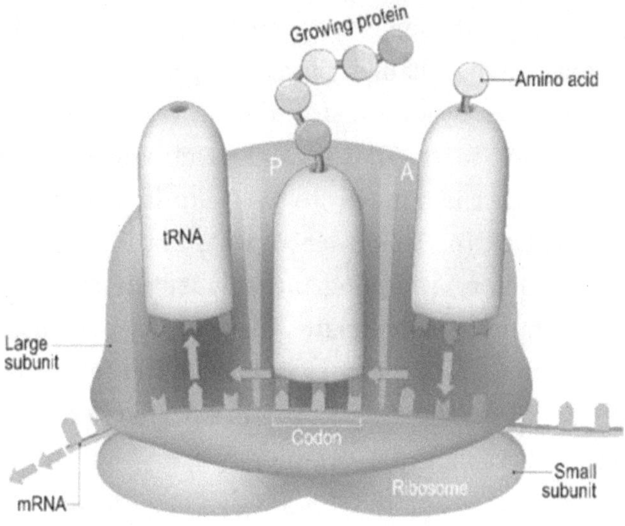

Source: Alamy. Reference OY0111316807

Figure 7: Biological protein synthesis- mRNA translation at a ribosome.

Ribosomes, molecular machines composed of ribosomal RNA (rRNA) and proteins, act as the sites of protein synthesis. Transfer RNA (tRNA) molecules serve as adaptors: each carries a specific amino acid and contains an anticodon that pairs with a matching mRNA codon. As the ribosome moves along the mRNA, tRNAs deliver their amino acids in the correct sequence. The ribosome then catalyzes the formation of a peptide bond between the incoming amino acid and the growing polypeptide chain. This process continues until a stop codon appears, signaling the end of translation and releasing the completed polypeptide. The translation cycle is highly coordinated. An mRNA molecule binds to the ribosome, and the small subunit scans until it finds the start codon. The first tRNA, carrying methionine, binds at the P site. The next tRNA enters the A site, and the peptidyl transferase activity of the large ribosomal subunit forms a peptide bond, transferring the growing chain to the amino acid in the A site. The ribosome then shifts by one codon during translocation. The empty tRNA moves to the E site and exits, while the tRNA carrying the growing chain moves to the P site, leaving the A site open for the next tRNA. This cycle repeats until a stop codon recruits release factors, releasing the polypeptide and causing the ribosomal subunits to dissociate. The accuracy of translation is remarkable, although rare errors can occur and may produce altered proteins. Understanding a protein's primary structure is like reading the first line of a complex recipe: it identifies the ingredients and their order, but not yet the final folded product. Even so, this sequence is essential for everything that follows. The chemical properties of amino acid side chains influence the forces driving protein folding, including hydrogen bonds, ionic interactions, hydrophobic forces, and van der Waals interactions. Therefore, primary structure is the foundation of the protein's final form and biological function, bridging genetic information in DNA to the dynamic proteins that carry out life's processes.

The number of amino acids in a polypeptide chain varies dramatically, from a few dozen to many thousands. Insulin, a relatively small protein hormone, consists of two polypeptide chains, one with 30 amino acids and the other with 21. By contrast, titin, also known as connectin, is a giant muscle protein containing more than 34,000 amino acids. An even larger protein, PKZILLA-1, was discovered in the marine alga *Prymnesium parvum*, or golden algae, and will be discussed in Chapter 5. This enormous range in protein length, combined with the diversity of the twenty amino acid building blocks, produces an almost limitless number of possible primary structures. On average, a protein is roughly one ten-thousandth the width of a human hair, meaning that about 10,000 proteins could fit side by side across a single hair.

This vast potential for variation enables life to generate the immense array of proteins required within a single organism and across the full diversity of life on Earth.

The genetic code, which determines the primary structure of proteins, is a remarkable evolutionary achievement. It is nearly universal across known life forms, from bacteria to mammals, suggesting that all life shares a common evolutionary origin. The code is also degenerate, meaning that more than one codon can specify the same amino acid. For example, six codons encode leucine. This degeneracy provides partial protection against mutations because a single nucleotide change may not alter the amino acid sequence. However, when a mutation does cause an amino acid substitution, the effects can be severe. In hypertrophic cardiomyopathy, for example, arginine at position 403 in MYH7 is replaced by glutamine, altering force generation in cardiac muscle. In long QT syndrome, single mutations in ion channels, such as KCNQ1, can disrupt electrical signaling in the heart. In osteogenesis imperfecta, replacement of a glycine in collagen often destabilizes the collagen triple helix because glycine is required at every third position.

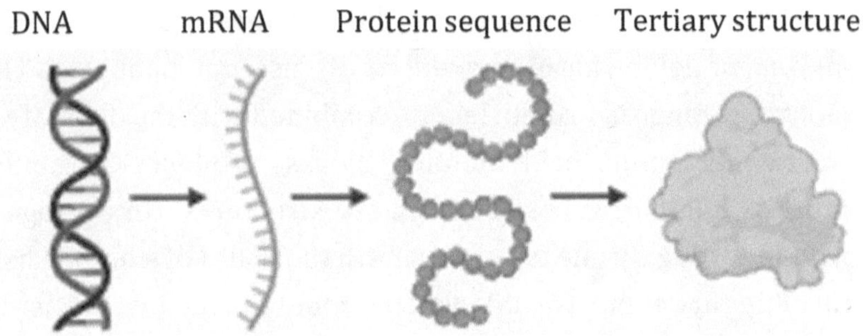

Figure 8. The central dogma of molecular biology: from DNA to proteins.

This classic view of gene expression was significantly expanded by Thomas Cech's work, which changed how scientists think about RNA and protein diversity. In the early 1980s, Cech proved that RNA is not just a passive intermediary between DNA and proteins but can also have catalytic activity. He showed that certain RNA molecules can cut and rearrange themselves without help from proteins, which had long been seen as the only biological catalysts capable of such complex tasks. Normally, genes are transcribed into RNA that contains introns, or noncoding regions, and exons, the coding regions. Before RNA can be used for protein synthesis, introns must be removed, and exons joined through splicing, usually done by protein-based machinery. However, some RNA molecules can perform this process themselves: they recognize and remove their own introns and correctly join their exons. Because they can both store genetic information and catalyze chemical reactions, these self-splicing RNAs are called ribozymes. This discovery overturned the idea that all biological catalysis is carried out solely by proteins. Along with Sidney Altman's independent work, it established RNA as an active participant in cellular chemistry and earned them the 1989 Nobel Prize in Chemistry. The implications were significant. First, the discovery showed that RNA processing is a dynamic and regulated step in gene expression. Mechanisms like RNA splicing permit precursor mRNA to be modified before translation. In more complex organisms, alternative splicing joins different combinations of exons, allowing a single gene to produce multiple mRNA transcripts and, therefore, multiple proteins with different functions. This greatly enhances proteomic diversity without increasing genome size. Cech's discovery also reinforced the idea of an early RNA-based form of life, often called the RNA world, in which RNA may have both stored genetic information and catalyzed biochemical reactions before the evolution of DNA and proteins.

In summary, although mRNA serves as the template for protein synthesis, the discovery of catalytic RNA showed that RNA itself plays an active and versatile role in gene expression. This insight reshaped molecular biology and helped explain how complex organisms can generate many proteins from a limited number of genes.

Even small changes in the primary sequence can have major consequences. Sickle cell anemia provides a classic example: replacement of glutamic acid with valine at position 6 in the beta-globin chain causes hemoglobin molecules to behave abnormally. Even seemingly conservative substitutions can have subtle effects. For example, replacing leucine with isoleucine preserves hydrophobicity, and the protein may still fold and function reasonably well. However, the slightly different size and shape of isoleucine can alter stability or interactions with other molecules. Over evolutionary time, such small differences can accumulate, generating variation in protein function and adaptation to different environments. For this reason, primary structure is central to understanding protein evolution. By comparing amino acid sequences of homologous proteins, proteins derived from a common ancestor, scientists can infer evolutionary relationships. Proteins essential for core cellular processes, such as DNA replication or metabolism, usually evolve slowly and remain highly conserved across organisms. For instance, the acetylcholine receptors found in sea rays (*Torpedo californica*, *Torpedo nobiliana*, and *Torpedo marmorata*) are remarkably similar to those in humans despite the evolutionary distance between these species. By contrast, proteins involved in more specialized functions or under weaker evolutionary pressure may accumulate mutations more rapidly. This comparative analysis helps build phylogenetic trees that depict the evolutionary history of life.

Understanding primary structure is also critical for protein engineering and drug design. If the amino acid sequence of a disease-related protein is known, researchers can design small molecules or antibodies that bind to it and modulate its activity. For example, if an enzyme is overactive in a disease, a drug may be designed to imitate the substrate without undergoing processing, thereby inhibiting the enzyme. Alternatively, a drug may bind to an allosteric site, alter the enzyme's conformation, and reduce its catalytic efficiency. These therapeutic strategies depend on understanding how primary sequence shapes three-dimensional structure and function.

The linear amino acid sequence forms the basis of protein chemistry and biology, making it one of the most crucial concepts for understanding life at the molecular level. Although it may seem simple, this sequence acts as a powerful carrier of genetic information, converting DNA into the molecules that keep life going. However, biological function rarely depends on a single string. Polypeptide chains must bend, fold, and arrange themselves into specific three-dimensional shapes to carry out their various roles. This folding is not random. Guided by the chemical properties of amino acids and the surrounding environment, the chain adopts recurring local structures that serve as building blocks for the entire protein architecture. These organized patterns include the α-helix and β-sheet, discussed in Chapter 2 (see Figure 3). They provide proteins with rigidity and a predictable shape, much like columns support a roof or woven threads reinforce fabric.

Let's first explore the α-helix. Think of a spiral staircase where each step is a specific distance from the next, winding upward in a consistent direction. In this structure, the polypeptide backbone coils into a right-handed helix. Its stability mainly comes from hydrogen bonds. Amino acids have an amino group (-NH2) and a carboxyl group (-COOH) attached to a central alpha carbon (see Figure 1). In a polypeptide chain, hydrogen bonds can form between backbone atoms. In the α-helix, the backbone carbonyl oxygen (C=O) of residue i bonds with the backbone amide hydrogen (N-H) of residue i+4; an α-helix contains about 3.6 amino acids per turn. This regular pattern creates a tightly coiled, rod-like structure (Figure 9). The side chains, or R groups, extend outward from the helical axis, like handrails protruding from a spiral staircase. This outward direction reduces steric hindrance and allows tighter packing. The R groups do not form the hydrogen bonds that define the helix, but their chemical properties strongly influence helix formation. Bulky or charged side chains can destabilize a helix, while smaller, nonpolar side chains often promote helix formation.

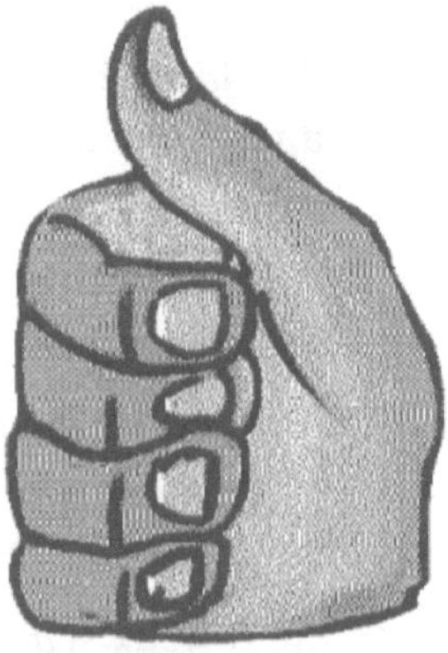

Figure 9. Right-hand α-helix model. In a right-handed coordinate system, when the thumb of the right hand points up, the curled fingers point from left to right, so it is a right-handed helix.

The α-helix is a common structural motif in proteins, found in everything from globular enzymes to structural proteins like keratin in hair and nails. Its compact, stable shape is ideal for regions that cross membranes, provide rigid support, or create interaction surfaces. Transmembrane proteins often contain long α-helices that pass through the lipid bilayer. The hydrophobic side chains in these helices interact favorably with lipid tails, anchoring the protein within the membrane. In other proteins, α-helices form coiled coils, where two or more helices twist around each other like ropes. This stable arrangement is seen in proteins such as tropomyosin, which helps muscle contraction. The predictable structure of the α-helix makes it a fundamental building block in protein architecture.

The β-sheet is a distinct yet equally important structural motif. Instead of forming a coil, the polypeptide extends into relatively flattened strands that align side by side (see Figure 3). A pleated sheet of paper, with folds creating ridges and valleys, offers a helpful analogy. Each individual strand is called a β-strand. Unlike α-helices, which form within a single continuous segment, β-sheets result from the association of multiple β-strands. These strands may originate from different regions of the same polypeptide or from separate chains. The primary stabilizing interactions are hydrogen bonds between backbone atoms of neighboring strands. Specifically, the N-H group in one strand forms a hydrogen bond with the C=O group in an adjacent strand. β-strands can align in two main orientations: parallel or antiparallel. In an antiparallel β-sheet, adjacent strands run in opposite directions (Figure 10): one from N-terminus to C-terminus and the neighboring strand from C-terminus to N-terminus.

Figure 10. Schematic representation of an anti-parallel β-sheet.

This arrangement produces relatively straight, stable hydrogen bonds oriented perpendicular to the strands. In a parallel β-sheet, adjacent strands run in the same direction, from N-terminus to C-terminus, resulting in slightly angled and somewhat less stable hydrogen bonds. Despite this difference, both parallel and antiparallel β-sheets are robust structural elements. The side chains in a β-sheet project alternately above and below the plane of the sheet, creating its characteristic pleated appearance. This arrangement allows close packing of polypeptide chains and contributes to structural stability.

β-sheets are often described as the walls or platforms of a protein. They provide a rigid, extended framework onto which other structural elements, such as α-helices, can attach or integrate. They are especially useful in the cores of globular proteins, where hydrophobic side chains are buried away from water while polar or charged side chains remain exposed on the surface. This hydrophobic effect is a major driving force in protein folding, and β-sheets often help organize it. Antibodies, for example, contain cores built from β-sheets arranged in sandwich-like structures that provide stable scaffolds for antigen-binding sites. Many enzymes also contain β-sheet cores that position catalytic residues for function.

The interplay among α-helices, β-sheets, and less regular structures, such as loops and coils, gives rise to the diverse three-dimensional forms of proteins (Figure 11).

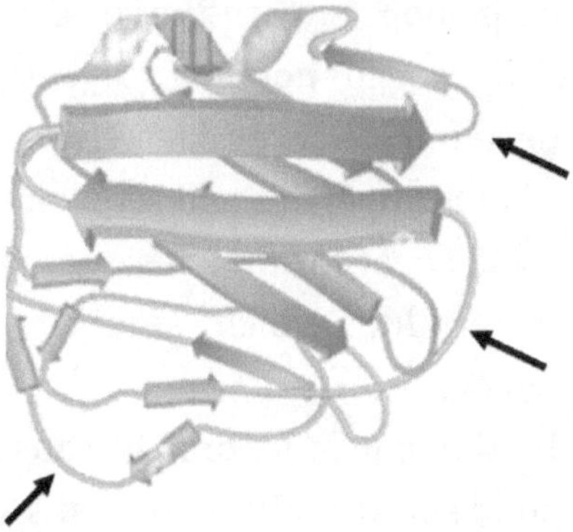

Figure 11. Tertiary structure of a protein. The arrows indicate the loops connecting α-helix and β-sheet structures.

These secondary structural elements do not form in isolation. They are connected by flexible loop regions, also called turns or coils. Although loops lack the regular hydrogen-bonding patterns of helices and sheets, they are essential. They allow the polypeptide chain to change direction, connect distant parts of the molecule, and position specific amino acid side chains for interactions with other molecules. Loops are often enriched in amino acids that are small, flexible, or capable of disrupting regular secondary structure.

In essence, α-helices and β-sheets are the main structural beams and supports, while loops act as connectors that let proteins bend, twist, and create specialized functional surfaces. Their widespread presence isn't random. These secondary structures are energy-efficient arrangements of the polypeptide backbone in water. Regular hydrogen bonding stabilizes these shapes, but the amino acid sequence decides where they form. Alanine and leucine, for example, often favor helix formation, whereas proline frequently appears in loops or turns because its cyclic structure disrupts regular hydrogen bonding. Glycine, being small and flexible, can also destabilize helices. Likewise, β-sheet formation depends on side-chain properties and sequence context. So, the placement of helix- and sheet-forming residues in the primary sequence helps determine which secondary structures develop.

The α-helix and β-sheet can be thought of as prefabricated modules. An α-helix is a tightly wound spiral, while a β-sheet is a pleated panel. These stable modules assemble in predictable ways, and loops connect them into a flexible, three-dimensional architecture (Figure 11). A protein's final shape depends on how these modules form, connect, and arrange themselves in space.

The discovery and characterization of secondary structures were pivotal moments in protein science. In the 1950s, Linus Pauling and Robert Corey proposed the α-helix and β-sheet based on theoretical calculations and analysis of peptide-bond geometry. Their work laid the foundation for understanding how a linear chain can fold into regular, predictable shapes. Later experimental studies, especially X-ray crystallography, confirmed their predictions and revealed how common these structures are in proteins. Visualizing them at atomic resolution has been essential for understanding their roles in protein function.

Secondary structural elements do more than maintain shape; they often contribute directly to function. In enzymes, the active site is frequently formed by residues from distant parts of the sequence brought together by folding. These residues may lie within α-helices, β-sheets, or connecting loops. Their precise spatial arrangement, shaped by secondary and tertiary structure, is essential for catalysis. In proteins that bind other molecules, such as DNA-binding proteins or antibodies, surfaces formed by α-helices and β-sheets often participate in recognition. Understanding how amino acid sequences direct the formation of helices and sheets allows scientists to design proteins with desired structural features and even create new protein-based materials. Researchers have engineered proteins that self-assemble into nanostructures by exploiting these predictable folding motifs. Thus, α-helices and β-sheets are fundamental recurring patterns of local folding. Stabilized by backbone hydrogen bonds, they provide order, rigidity, and consistent geometry, forming the foundation for the more complex tertiary and quaternary structures that follow.

The process of transforming a linear amino acid chain into a functional protein involves multiple levels of organization. We have explored the primary sequence, which serves as the essential blueprint, and the secondary structures, such as α-helices and β-sheets, that form recurring local patterns. Next, we examine the tertiary structure. This stage involves a polypeptide chain, with its secondary structural elements and amino acid side chains, folding into a unique, three-dimensional, and often compact functional shape. Imagine a sculptor shaping soft clay. The sculptor first creates repeating patterns and then twists, bends, and folds those sections into a final form. In proteins, this sculpting is guided by the chemical properties of amino acid side chains and the forces that influence their interactions. The complete three-dimensional shape of a single polypeptide chain is its tertiary structure.

TERTIARY STRUCTURE → ***FUNCTIONAL SHAPE***

Scheme 1

Every twist, bend, and contact between distant parts of the polypeptide chain is influenced by the chemical nature of amino acid side chains. These interactions provide the glue and scaffolding that maintain the protein's functional shape. Noncovalent interactions and, in some cases, covalent bonds act together to stabilize the protein's architecture. Their collective effect, encoded by the amino acid sequence, drives the chain toward its unique tertiary structure.

One of the main driving forces in forming tertiary structures is the hydrophobic effect. Amino acid side chains can be polar, nonpolar, or charged. In the watery environment of the cell, polar water molecules interact favorably with other polar or charged groups. Nonpolar, hydrophobic side chains, however, cannot form favorable hydrogen-bonding interactions with water. Instead of imagining hydrophobic groups as actively sticking together, it is more accurate to view the process from water's perspective: water strongly prefers to keep its hydrogen-bonded network intact. Hydrophobic groups disrupt this network, so water effectively excludes them. The same principle is seen when oil is added to water or vinegar: the oil forms droplets, not because oil molecules strongly attract each other, but because water reorganizes to preserve its own hydrogen bonding. In proteins, hydrophobic side chains are pushed away from water and cluster in the interior, forming a stable hydrophobic core. Solvent-exposed loop regions, on the other hand, are enriched with polar, charged, and flexible residues that interact well with water. This setup shields hydrophobic residues from the aqueous environment and is a powerful thermodynamic driver of folding. As a result, many globular proteins have hydrophobic interiors and more hydrophilic surfaces. The primary sequence determines where these nonpolar clusters form and helps guide the overall folding process.

Hydrogen bonds complement the hydrophobic effect and help stabilize the tertiary structure. Besides backbone-to-backbone hydrogen bonds that define α-helices and β-sheets, hydrogen bonds can form between polar side chains or between a side chain and the polypeptide backbone. The hydroxyl groups of serine or threonine and the amide groups of asparagine or glutamine, for example, can participate in these interactions. Although individual hydrogen bonds are weaker than covalent bonds, many such bonds collectively help stabilize a protein's specific fold. Ionic bonds, or electrostatic interactions, add another layer of stability. They form between side chains with opposite charges, such as negatively charged aspartate or glutamate and positively charged lysine, arginine, or histidine. These interactions, often called salt bridges, can strongly stabilize the tertiary structure, especially when buried away from water. They act like molecular magnets, drawing distant regions of the chain together.

Van der Waals forces also contribute to precise protein packing. These weak, short-range attractions arise from temporary fluctuations in electron distribution around atoms. Individually, they are subtle, but when many atoms are packed closely together in a folded protein, their cumulative effect can be substantial. They help structural elements fit tightly and contribute to overall stability.

A particularly important covalent bond that can influence tertiary structure is the disulfide bridge, introduced in Chapter 2. Disulfide bridges form when the sulfhydryl (-SH) groups of two cysteine residues oxidize and create a covalent sulfur-sulfur bond (-S-S-). This bond links different parts of the polypeptide, creating a strong bond that enhances stability. Disulfide bridges are especially common in secreted proteins and in proteins that function outside the cell, where they provide added resistance to denaturation. The formation of these bridges is often a key step in protein maturation. Together, hydrophobic interactions, hydrogen bonds, ionic bonds, van der Waals forces, and disulfide bridges determine the three-dimensional arrangement of the polypeptide chain. Folding is a complex process, yet it is driven by information encoded in the amino acid sequence. The protein does not require an external instruction manual; the instructions are built into its sequence. Tertiary structure is the final molecular sculpture, in which primary and secondary structural elements and side chains are arranged in space to create a unique functional form. Every curve, angle, and surface feature contributes to the protein's function, much like a key that fits a lock.

The arrangement of amino acid side chains in the tertiary structure creates binding sites, active sites, and surfaces that interact with other molecules. Tertiary structure also depends on the protein's environment. The forces that stabilize folding are sensitive to pH, temperature, mechanical stress, and salt concentration. Changes in these conditions can disrupt the noncovalent interactions that maintain the folded state, leading to denaturation. Increased temperature, for example, raises molecular motion and can break hydrogen bonds and van der Waals interactions. Extreme pH can alter the charge states of amino acid side chains and disrupt ionic bonds. Denaturation usually causes loss of function, emphasizing the importance of the precise three-dimensional fold. A familiar example of temperature-induced protein change is frying an egg, as discussed in Chapter 1. Acidic conditions can also alter proteins, as seen in the formation of yogurt and cheese. Now consider myoglobin, an oxygen-storing protein found in muscle cells. Myoglobin has a compact, globular tertiary structure composed largely of α-helices. Embedded within it is a heme group, a complex molecule that binds oxygen. The precise placement of the heme group within the myoglobin pocket, maintained by the tertiary fold and by interactions with specific amino acid side chains, is essential for efficient oxygen binding and release. If myoglobin unfolds and loses its tertiary structure, it can no longer bind oxygen effectively.

Hemoglobin, the oxygen-transport protein in red blood cells, provides another example. It is composed of four separate polypeptide chains, each with its own tertiary structure (see Figure 4). These four chains assemble into a quaternary structure, discussed later in this chapter. The tertiary structure of each hemoglobin subunit is essential for oxygen binding and for the cooperative behavior that allows hemoglobin to take up oxygen in the lungs and deliver it to tissues efficiently.

The Tertiary Structure is Unique to Each Protein

The study of tertiary structure relies on experimental techniques that reveal the three-dimensional arrangement of atoms within proteins. X-ray crystallography and nuclear magnetic resonance

(NMR) spectroscopy are two major methods. X-ray crystallography requires crystallizing a protein and exposing it to X-rays; the resulting diffraction pattern provides information about electron density, from which a three-dimensional model can be built. NMR spectroscopy determines protein structure by analyzing how atomic nuclei, especially hydrogen atoms, behave in a strong magnetic field. These nuclei absorb and re-emit radiofrequency signals that depend on their chemical environment and spatial relationships. By measuring interactions between nearby atoms, scientists infer distances and angles, which are then used computationally to generate structural models. These methods have transformed our understanding of protein structure and function, revealing the atomic details of many tertiary structures. The beauty of tertiary structure lies in its diversity and specificity. While secondary structures recur across proteins, each tertiary structure reflects a particular amino acid sequence and enables a particular function. This uniqueness allows proteins to catalyze reactions, transport molecules, provide support, and mediate signaling. Tertiary structure is the final three-dimensional form adopted by a single polypeptide chain to achieve biological function, showing the elegance of molecular self-assembly. This concept is especially important in biopharmaceutical development, where therapeutic proteins must maintain their folded structures to remain safe and effective. A major goal of formulation science is to stabilize the tertiary structure during storage by controlling temperature, pH, and the use of stabilizing additives. Preventing denaturation helps ensure that therapeutic proteins retain their intended function when administered to patients.

Quaternary Structure

The three-dimensional structure of a single polypeptide chain, called its tertiary structure, is a remarkable achievement in molecular self-assembly. However, for many proteins, this is not the final level of organization. Quaternary structure explains how multiple folded polypeptide chains come together to form a larger, functional protein complex, as shown for hemoglobin (see Figure 4). Not all proteins need multiple subunits; many work as single chains. But for many biological functions, assembling into a multi-subunit structure is essential. Imagine several finely crafted sculptures fitting together precisely to create a larger form. In proteins, these sculptures are subunits, and their arrangement determines the quaternary structure. Subunits are held together by interactions similar to those stabilizing tertiary structures: hydrogen bonds, ionic bonds, van der Waals forces, and hydrophobic interactions. Sometimes, disulfide bonds also connect subunits. The number, strength, and arrangement of these interactions influence the stability and shape of the protein complex. Subunits can assemble in various ways, forming dimers, trimers, tetramers, rings, linear chains, or larger oligomers such as the proteasome, which breaks down damaged or unnecessary proteins. These assemblies may display rotational, helical, or more complex symmetries.

Advantages of Multisubunit Complexes

The assembly of multiple subunits offers several benefits that help explain the evolution of quaternary structure. One major benefit is cooperativity. Hemoglobin, which carries oxygen in blood, provides the classic example. A single hemoglobin molecule contains four polypeptide subunits: two alpha (α) globin chains and two beta (β) globin chains (Figure 4). Each subunit can bind one oxygen molecule. Importantly, binding the first oxygen molecule causes a conformational change that transmits through subunit interfaces, increasing the oxygen affinity of the remaining subunits. Multisubunit assembly can also create binding sites that individual subunits do not have. Interfaces between chains may form pockets or grooves designed to bind specific ligands, substrates, DNA, or other proteins. These interface sites are often crucial for function. The ribosome, the cellular machinery responsible for protein synthesis, demonstrates this principle on a large scale. It consists of ribosomal RNAs and dozens of proteins arranged into an organized assembly that enables accurate and efficient translation of genetic information (see Figure 7). The study of quaternary structure often overlaps with that of tertiary structure, as each subunit must fold properly before proper assembly can take place. X-ray crystallography and cryo-electron microscopy are particularly useful for visualizing subunit organization. Cryo-EM, in particular, has revolutionized the study of large protein complexes by revealing detailed structures without requiring crystallization.

Interactions at subunit interfaces are highly specific and can be regulated by other molecules. The binding of a small molecule or protein to one subunit may change the shape of the whole complex and influence its activity. This allosteric regulation, which works through quaternary structure, is common in cellular signaling and metabolic control.

The ability of proteins to form multisubunit complexes demonstrates the power of modular design in biology. Before moving on, it is useful to distinguish quaternary structure from self-association. Quaternary structure refers to a defined, biologically functional assembly of subunits. Protein self-association is broader and describes the tendency of protein molecules to interact with one another, either temporarily or permanently. Such interactions may produce normal complexes, larger clusters, aggregates, or fibers, none of which necessarily serve a biological purpose. By dividing complex tasks among smaller subunits that can be made and assembled independently, cells build molecular machines with remarkable efficiency and adaptability. Quaternary organization allows proteins to perform functions and regulatory tasks that individual chains alone could not achieve, illustrating the principle that the whole can be greater than the sum of its parts.

Structure and Function

The principle that structure determines function is fundamental to protein biology, although important exceptions will be discussed later in the book. Each twist, turn, and interaction within a protein molecule is coordinated to support a specific cellular or organismal task. To understand how proteins work, and how they fail, we must understand the connection between form and function. Consider enzymes, the workhorses of metabolism. Their catalytic power does not come solely from the amino acid chain but from the precise three-dimensional arrangement of atoms in the active site. The shape, charge distribution, and chemical properties of amino acids lining the active site complement those of the substrate. This fit is often compared to a lock and key, although many enzymes adjust their shape upon binding in an induced-fit mechanism. If the protein structure changes, even slightly, the active site may distort, preventing substrate binding or disrupting catalysis. The result can be a stalled biochemical pathway with cascading effects on the cell. The principle extends well beyond enzymes. Transport proteins, such as ion channels and carrier proteins embedded in cell membranes, depend on precise conformational changes to move molecules and ions across the otherwise impermeable lipid bilayer. Receptors also rely on their folded structures to detect signals and convert binding events into cellular responses.

Structural proteins like collagen and keratin demonstrate the relationship between structure and function on a larger scale. Their organized fibrous structures give them strength and resistance to stretching. Mutations in collagen genes can lead to connective tissue disorders, including Ehlers-Danlos syndrome and osteogenesis imperfecta. In these conditions, defective collagen weakens tissues, resulting in joint hypermobility, fragile skin, skeletal abnormalities, or brittle bones. Collagen's strength depends not only on individual molecules but also on their organized assembly into fibrils and fibers, showing that structure at various levels influences function. DNA- and RNA-binding proteins also depend on specific structural motifs to recognize their targets. Helices, loops, and other motifs fit into DNA grooves or contact the sugar-phosphate backbone, with particular amino acid side chains forming hydrogen bonds and van der Waals interactions with nucleic acid bases. A misplaced amino acid or slight folding defect can disrupt binding and cause errors in gene regulation. Similarly, accurate DNA replication relies on the structural compatibility among the DNA template, incoming nucleotides, and the DNA polymerase active site. Defects in polymerase structure can increase errors and contribute to diseases such as cancer.

Protein structure is also flexible. Although proteins are often depicted as rigid three-dimensional objects, they are dynamic molecules in constant movement. Many experience subtle shape changes, or conformational fluctuations, sometimes called breathing motions. These motions range from quick atomic vibrations to larger shifts of entire protein regions. They can open and close channels, reveal hidden binding sites, regulate interactions, and enable proteins to respond to environmental changes. In many cases, function depends not on one fixed structure but on the ability to switch among related conformations. During muscle contraction, actin and myosin repeatedly change shape to produce force. In signaling, receptors change conformation when hormones, neurotransmitters, or other molecules bind, triggering downstream events. Enzymes also depend on flexibility; in the induced-fit model, substrate binding adjusts the active site to improve catalytic alignment and efficiency.

Understanding protein motion is increasingly vital in drug discovery, where medicines may be designed to stabilize or disrupt specific conformations. Proteins achieve dynamic movement through a balance of stabilizing and destabilizing forces. Genetic mutations or environmental changes can disrupt this balance, locking a protein into an inactive or overactive state. The cellular environment itself also impacts protein structure and function. As mentioned earlier, pH, temperature, shear stress, and other conditions influence stability. Extreme conditions can cause denaturation, where a protein unfolds and loses its specific three-dimensional structure, usually resulting in a loss of function. This is why the body tightly controls temperature and pH. When these conditions are disturbed, as in fever or metabolic disease, protein function can be compromised.

Structural Understanding

Protein misfolding diseases provide some of the strongest evidence for the structure-function relationship. Alzheimer's disease, Parkinson's disease, and cystic fibrosis are all connected to proteins that fail to fold properly. These conditions demonstrate that when protein structure is compromised, function can be lost, and the consequences can be severe. The underlying cause may be a subtle change in amino acid sequence, even a single mutation, which disrupts folding and results in a structurally flawed protein. Understanding three-dimensional protein structure is therefore not only an academic pursuit but also a fundamental aspect of modern molecular biology and medicine. Structural information reveals how proteins function at the molecular level: how enzymes catalyze reactions, how binding sites recognize targets, how conformational changes occur, and how proteins interact with each other.

This structural understanding is essential for rational drug design. When researchers determine the three-dimensional structure of a disease-related protein, they can create small molecules or antibodies that bind to and control its activity. If a protein is overactive, a drug may be designed to block its active site or bind to an allosteric site that stabilizes an inactive form. If a protein's activity is decreased, another drug might stabilize its active form or improve substrate binding. Many modern drugs, including anticancer kinase inhibitors and antivirals targeting viral enzymes, rely on our ability to understand and modify protein structures. The structure-function paradigm also underpins synthetic biology and protein engineering. Scientists can now design new protein sequences predicted to fold into specific shapes with particular functions, allowing for custom enzymes, biosensors, and innovative protein-based therapies.

CHAPTER 4
The Dance of Folding and Misfolding

The journey from a linear string of amino acids to a functional three-dimensional protein is one of biology's most elegant and fundamental processes. For many proteins, folding occurs with remarkable speed, often within microseconds to milliseconds. This rapid formation of the correct structure is essential because proteins must become functional soon after they are made. Folding is a spontaneous act of molecular self-assembly, guided not by chance but by physical and chemical forces encoded in the amino acid sequence. At the heart of this process is thermodynamics. Like all molecules, proteins tend to move toward lower-energy states. The unfolded polypeptide chain is relatively high in energy and disorder, whereas the folded three-dimensional structure is usually lower in energy and more stable. The amino acid sequence, determined by the genetic code, serves as a blueprint that guides the protein toward its most stable and functional form.

One of the main forces driving protein folding is the hydrophobic effect, as discussed in earlier chapters. Water, which makes up most of the cellular environment, is polar and interacts well with charged or polar groups. Nonpolar molecules, however, do not interact effectively with water. As a polypeptide begins to fold, hydrophobic amino acids tend to cluster in the interior, shielding themselves from the surrounding aqueous environment, while hydrophilic regions remain exposed. This uneven distribution of polar and nonpolar residues influences the energy landscape that guides folding. In other words, water-water interactions are favored over water-hydrocarbon interactions, and this preference drives hydrophobic regions into the protein's interior.

Beyond the hydrophobic effect, several other forces help shape proteins accurately. Hydrogen bonds are especially important. These weak but numerous electrostatic attractions form between a hydrogen atom covalently bonded to an electronegative atom, such as oxygen or nitrogen, and another nearby electronegative atom. In proteins, hydrogen bonds can form between backbone atoms, stabilizing secondary structures such as α-helices and β-sheets, and between amino acid side chains, helping refine tertiary structure. Although each hydrogen bond is weak, its combined effect strongly stabilizes the final folded shape.

Ionic bonds, also called salt bridges, and van der Waals interactions further stabilize folded proteins. Salt bridges form between oppositely charged side chains, such as positively charged lysine or arginine and negatively charged aspartate or glutamate. These electrostatic attractions help keep distant parts of a protein together, especially in the protein core or at interfaces. Van der Waals forces are weak, short-range attractions caused by temporary fluctuations in electron distribution around atoms. Although small on their own, they can add up significantly when many atoms are tightly packed within a folded protein, particularly in the hydrophobic core. In some proteins, disulfide bonds provide covalent stabilization. Together, the hydrophobic effect, hydrogen bonds, ionic bonds, van der Waals forces, and disulfide bonds form a complex energy landscape that guides the polypeptide chain toward its stable, functional shape.

Many proteins transform from linear chains into functional three-dimensional structures within milliseconds or seconds. This speed is remarkable given the enormous number of possible conformations available to a flexible polypeptide chain. Proteins fold quickly and reliably because the process is guided by the energy landscape rather than by random search. This precision is essential: even small deviations from the proper fold can reduce function or create harmful new properties.

Levinthal's Paradox

Levinthal's paradox highlights the extraordinary efficiency of protein folding. In 1969, biophysicist Cyrus Levinthal observed that if a newly made protein randomly sampled every possible shape to find its correct three-dimensional structure, the search would take far longer than the age of the universe. Yet experiments show that many proteins fold into functional forms within milliseconds to seconds. The paradox arises because proteins are long chains of amino acids linked by flexible bonds, and each bond can rotate in multiple ways, creating an immense number of possible conformations. Even small proteins can theoretically adopt trillions of shapes. If folding were purely random, finding the correct structure would be nearly impossible. Proteins avoid this problem by following organized, directed pathways. As a protein emerges from the cellular machinery that synthesizes it, interactions among amino acids begin guiding local folding. Regions quickly form structures such as α-helices and β-sheets, which then interact to create more complex arrangements. This process dramatically reduces the number of conformations that must be explored.

The solution to Levinthal's paradox is that folding occurs along specific pathways on an energy landscape, often described as a "folding funnel" (Figure 12). Rather than wandering randomly through every possible shape, the protein is biased toward lower-energy conformations. At each step, chemical and physical forces—including hydrogen bonding, hydrophobic interactions, electrostatic attractions, van der Waals forces, and disulfide-bond formation—stabilize favorable structures and discourage unfavorable ones.

A useful analogy is a ball rolling downhill. Rather than searching every square inch of a mountain range, the ball naturally moves toward lower elevations. Similarly, a protein is guided toward its native structure because that state is usually energetically favored. Levinthal's paradox, therefore, shows that protein folding is not a random trial-and-error process. It is an organized, stepwise process guided by the physical and chemical properties of amino acid sequences, enabling proteins to form functional structures rapidly and reliably within living cells.

Protein folding is dynamic and sometimes requires assistance. These helpers, called chaperones, are specialized proteins that assist other proteins in folding properly and help prevent aggregation and misfolding. Misfolded proteins can lose function and, in some cases, accumulate into toxic aggregates associated with diseases such as Alzheimer's, Parkinson's, and Huntington's disease. The study of protein folding, therefore, has profound implications not only for normal biology but also for understanding disease.

The Folding Funnel

The folding funnel concept illustrates how proteins move from many possible unfolded states toward a smaller set of stable folded states. Imagine a funnel that is wide at the top and narrow at the bottom. The top represents numerous high-energy unfolded conformations. As the protein folds, it moves down the funnel (Figure 12), exploring fewer conformations as it generally progresses toward lower-energy states. The bottom represents the native folded state, or a small set of closely related low-energy states, corresponding to the functional three-dimensional structure.

The funnel shape emphasizes that folding does not follow a single predetermined route. Proteins may take multiple pathways to reach the same final structure. Along the way, they can encounter small energy barriers or become temporarily trapped in local minima corresponding to partially folded intermediates. Nevertheless, the overall landscape biases the chain toward the global energy minimum, allowing most proteins to fold efficiently despite the enormous number of possible conformations.

The hydrophobic effect is a major driving force in this process. Hydrophobic amino acid side chains tend to cluster in the protein interior, away from water, causing the polypeptide chain to collapse into a compact structure. This initial collapse is followed by the formation and refinement of secondary and tertiary structures, stabilized by hydrogen bonds, van der Waals interactions, ionic interactions, and, in some proteins, disulfide bonds.

Efficient spontaneous folding is essential for life. If proteins required minutes or hours to fold, or if they often failed to reach their native conformations, newly synthesized proteins would accumulate faster than they could become functional. Cells would be overwhelmed by nonfunctional or misfolded proteins prone to aggregation, disrupting cellular organization and contributing to disease.

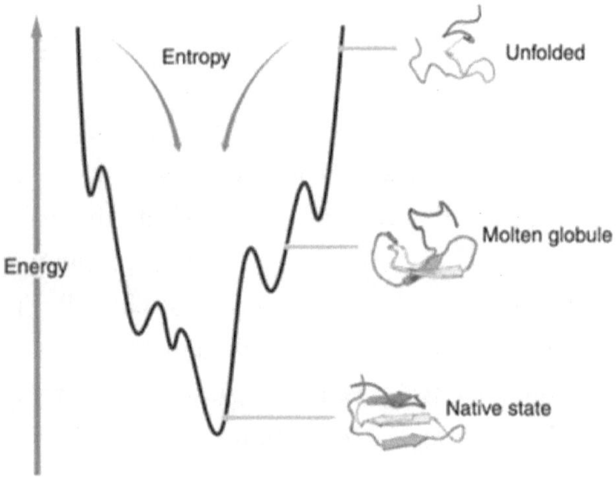

Figure 12: Folding funnel. Protein energy landscape.

The ability of proteins to find their correct three-dimensional shapes through self-assembly, guided by thermodynamics and balanced chemical forces, is fundamental to cellular function and ultimately to life. It illustrates how the laws of physics and chemistry, acting at the molecular level, produce the organized complexity of biology.

Inside the crowded cell, protein folding is both critical and precarious. Although amino acid sequences are the primary architects of protein structure, the cellular environment is far from a sterile laboratory. It is densely packed with molecules that collide, interact, and compete for space. This crowding can challenge the delicate, transient states of unfolding or partially folded proteins. In such an environment, unintended interactions can lead to misfolding or the formation of nonfunctional aggregates. Molecular chaperones, introduced above, act as essential guardians of protein integrity. They guide, protect, and assist proteins as they move from linear chains to correctly folded structures, especially during stress.

A useful analogy is a complex maze. The protein's amino acid sequence is the map, guiding the route to the functional state. However, the maze contains dead ends and confusing turns. Unfolded proteins may follow incorrect paths, become trapped, or collide with other vulnerable proteins, leading to tangles. Chaperones act like experienced guides: they do not rewrite the map, but they help keep proteins on productive pathways, prevent inappropriate interactions, and sometimes give a misfolded protein another chance to fold correctly. This intervention is crucial because once proteins aggregate, restoring them to their proper form can be difficult or impossible.

Molecular chaperones play several roles. Some act as holdases, temporarily binding unfolded or partially folded polypeptides. This prevents vulnerable proteins from interacting with one another and forming harmful aggregates. By sequestering newly made chains, holdases buy time, allowing proteins to explore their conformational landscape and reach their native states with less interference. This function is especially important for slowly synthesized proteins or proteins with flexible regions prone to aggregation.

Other chaperones act more actively as folding catalysts or refolding agents. They can bind misfolded proteins and, through cycles of binding and release, help unfold misfolded regions so the protein can try again. This unfolding-refolding cycle gives proteins renewed access to productive folding pathways and is especially important for correcting errors caused by initial misfolding or cellular stress.

Chaperone specificity is an active area of research. Some chaperones bind broadly to exposed hydrophobic patches on unfolded proteins, while others recognize specific sequence motifs or structural features. This selectivity helps match the right chaperone to the right substrate at the right time. Chaperones also coordinate folding with synthesis and degradation pathways. They can escort newly made proteins from the ribosome to their destinations, protecting them during transit. If a protein cannot be folded or refolded, it may be targeted for degradation by the cell's quality-control systems, preventing toxic buildup.

The network of chaperones and co-chaperones forms an advanced cellular quality-control system. It is essential not only for folding newly synthesized proteins but also for maintaining protein quality throughout their lifespan. As cells age or experience chronic stress, this system can become less effective, allowing damaged or misfolded proteins to accumulate. This decline contributes to aging and age-related disease.

The study of chaperones has revolutionized our understanding of cell biology and disease. Proteins do not always fold correctly on their own; specialized cellular machinery assists them in achieving and maintaining functional shapes. This discovery has created new therapeutic options, including efforts to boost chaperone activity in diseases marked by protein buildup. Chaperones are not passive helpers; they are active participants in cellular processes, integrated with synthesis, trafficking, stress responses, and degradation. Their significance becomes evident when folding goes wrong. Misfolding is not just a superficial issue. It can lead to loss of function, expose sticky surfaces, promote aggregation, and produce toxic species that interfere with cellular machinery.

The cellular environment is highly concentrated, with macromolecule levels estimated at roughly 300–400 mg/mL. This macromolecular crowding strongly affects protein behavior. Crowding can favor compact states, including partially folded or aggregated forms, over more extended conformations. Chaperones counteract this risk by binding exposed regions that might otherwise interact inappropriately or by providing protected spaces in which folding can proceed with less interference.

Heat Shock Helpers

When cells are exposed to higher temperatures, proteins become more likely to denature, and their three-dimensional structures can break down. This loss of structure often results in aggregation. In response, cells sharply increase the production of heat shock proteins (HSPs), a main group of chaperones. Hsp70 and Hsp90 are among the most well-studied examples. They assist in refolding denatured proteins and prevent toxic aggregates, acting like emergency responders that restore order during cellular stress. Other stresses, such as oxidative damage, toxins, and lack of nutrients, can also boost chaperone activity. This response helps keep protein homeostasis, which balances protein synthesis, folding, and degradation.

Many chaperones, particularly the Hsp70 family, operate through ATP-regulated cycles of binding and release. In the ATP-bound state, they bind exposed polypeptide segments. ATP hydrolysis to ADP changes the chaperone's shape and increases its grip on the substrate. Subsequent ATP binding promotes release, allowing the protein another opportunity to fold. Some chaperones work within larger multisubunit complexes called chaperonins. GroEL/GroES in bacteria and TRiC/CCT in eukaryotes form barrel-like structures with central cavities. An unfolded or misfolded protein enters the cavity, a lid closes, and the protein folds in relative isolation, protected from the crowded cellular environment.

When a protein misfolds, its functional structure is altered. Active sites or binding surfaces may become distorted, hidden, or misplaced, preventing the protein from performing its proper role. In a cell that depends on precise enzymatic activity, signaling, transport, and structural support, the loss of even a single crucial protein can have widespread effects. Misfolded proteins can also gain harmful properties. They may interfere with the folding of other proteins, disrupt cellular pathways, or become prone to aggregation.

In the crowded cellular environment, unfolded or partially folded proteins are constantly at risk of aberrant interactions. Instead of reaching their native shapes, they may collide, triggering a chain of misfolding events. Misfolded proteins can form amorphous clumps or highly organized amyloid-like fibrils. These aggregates are often stable and resistant to degradation, allowing them to accumulate inside or outside cells.

Protein aggregates can harm cellular structures, block transport pathways, and overwhelm quality-control systems. Similar to a traffic jam that halts traffic on a highway, aggregates can impede molecular traffic, disrupt organelle movement, and trigger inflammatory responses. Cells expend significant energy clearing these aggregates, but when production exceeds clearance capacity, damage accumulates. The effects of misfolding and aggregation extend beyond individual cells and contribute to many human diseases. While mechanisms differ depending on the protein, tissue, and aggregate type, the core issue is the disruption of normal cellular and tissue functions.

Diseases Linked to Protein Unfolding and Aggregation

Alzheimer's disease, a leading cause of dementia, demonstrates the role of protein aggregation in illness. It is linked to two main protein abnormalities: amyloid plaques and neurofibrillary tangles. Amyloid plaques develop outside neurons and mainly consist of amyloid-beta (Aβ), a peptide produced from amyloid precursor protein (APP). In healthy individuals, Aβ peptides are normally removed from the brain. In Alzheimer's disease, changes in APP processing or defective clearance can result in excess Aβ, which forms toxic oligomers and insoluble fibrils between neurons. These plaques can disrupt neuronal communication and trigger inflammation. Tau, another protein, normally stabilizes microtubules that support neuronal structure and intracellular transport. In Alzheimer's disease, tau becomes abnormally modified, separates from microtubules, and forms neurofibrillary tangles inside neurons. Together, Aβ plaques and tau tangles create a harmful environment that leads to neuronal dysfunction, memory loss, confusion, and ongoing cognitive decline. However, the relationship is complex: some older individuals without cognitive impairment also have beta-amyloid plaques, and why these deposits cause disease in some people remains unclear.

Parkinson's disease is another major neurodegenerative disorder linked to protein misfolding. It primarily affects motor control and is characterized by the loss of dopaminergic neurons in the substantia nigra. Central to this disease is the misfolding and aggregation of α-synuclein. In healthy neurons, α-synuclein is thought to help regulate neurotransmitter release. In Parkinson's disease, it forms aggregates known as Lewy bodies. These structures, found in affected neurons, may disrupt cellular transport, stress responses, and other essential processes, contributing to neuronal death. The progressive loss of dopaminergic neurons produces tremor, rigidity, bradykinesia, and postural instability. Lewy bodies can also appear in other brain regions and in the peripheral nervous system, contributing to cognitive and autonomic symptoms.

Aging is a major risk factor for many protein misfolding diseases. As organisms age, chaperone networks may weaken, protein-synthesis fidelity may decline, and waste-disposal systems such as the proteasome and autophagy may become less efficient. This gradual loss of protein homeostasis allows misfolded proteins to accumulate more readily, shifting the balance toward disease. Understanding aging is, therefore, closely linked to understanding proteinopathies.

Beyond the brain, protein misfolding contributes to systemic diseases. Cystic fibrosis illustrates how a single misfolded protein can have widespread effects. The most common mutation, ΔF508, affects the cystic fibrosis transmembrane conductance regulator

(CFTR), an ion channel. This mutation causes CFTR to misfold during synthesis in the endoplasmic reticulum. Instead of reaching the plasma membrane, the defective protein is recognized by quality-control systems and degraded. As a result, functional CFTR is absent or reduced at the cell surface, leading to abnormal ion and water transport across epithelial tissues. Thick, sticky mucus accumulates in organs such as the lungs and pancreas, causing chronic infections and digestive complications. Although cystic fibrosis does not involve large insoluble aggregates like many neurodegenerative diseases, its central problem remains defective protein folding and trafficking. In some cancers, abnormal folding or aggregation can also disrupt signaling, promote genomic instability, and influence drug resistance. Increased chaperone activity in cancer cells may help stabilize oncogenic proteins and support tumor survival.

Because proteins are dynamic and delicate, the risk of misfolding is always present. A single amino acid substitution can destabilize a native structure, and environmental factors such as oxidative stress, heavy metals, or viral infection can damage proteins and promote misfolding. Cells, therefore, rely on robust defense systems to maintain protein quality. Even small folding errors can have serious consequences.

The cell is a busy molecular environment, and when proteins stray from their proper folds, they may lose the ability to perform precise interactions. Misfolded proteins can also expose hydrophobic surfaces that are normally buried within the protein core. These exposed surfaces promote interactions with other misfolded proteins, leading to the formation of larger assemblies called aggregates. Protein aggregation is a visible sign of failed quality control and can have serious cellular consequences.

Imagine a well-organized library where books are properly cataloged and shelved. When a few books are damaged or misplaced, they become hard to use. If they start sticking together and forming piles in the aisles, they block access and disrupt the entire system. Something similar happens inside cells when misfolded proteins clump together. Usually soluble proteins start forming insoluble masses that can interfere with vital processes and contribute to disease.

The tendency of misfolded proteins to aggregate is closely linked to surface changes. Properly folded proteins display functional groups that support interactions with water, substrates, cofactors, or other proteins. When a protein misfolds, hydrophobic regions that are normally hidden in the core can become exposed. These exposed patches attract similar regions on other misfolded proteins, promoting self-assembly. The process becomes a chain reaction in which each new interaction helps grow and stabilize the aggregate.

Protein aggregates vary widely depending on the protein, cellular conditions, and degree of misfolding. Some are amorphous clusters that disrupt cellular space and function. Others form highly ordered amyloid fibrils with cross-β-sheet architecture, making them especially stable and resistant to degradation. Aggregates can develop in the cytoplasm, nucleus, endoplasmic reticulum, mitochondria, or extracellular space. They may block transport pathways, hinder organelle movement, deactivate enzymes or transcription factors, and interfere with chaperones. This creates a vicious cycle: aggregates damage the systems needed to remove them. Accumulation of aggregates can also trigger stress responses, such as the unfolded protein response in the endoplasmic reticulum. If stress is severe or persistent, these responses may cause cell death. In tissues, protein aggregates can also trigger inflammation, contributing to disease progression, especially in neurodegenerative disorders.

Different misfolded proteins can follow different aggregation pathways, producing aggregates with distinct structures, compositions, and cellular locations. This diversity helps explain why protein misfolding diseases vary widely in symptoms, affected tissues, and rates of progression.

Cells possess sophisticated quality-control systems to prevent or reduce aggregation, but these systems are not perfect. With age, chaperone networks may weaken, protein-synthesis accuracy may decline, and the ubiquitin-proteasome system and autophagy may become overwhelmed. This gradual loss of proteostasis creates conditions in which misfolded proteins accumulate more easily, increasing the risk of aggregation and disease.

Protein aggregation impacts many human diseases. Besides Alzheimer's and Parkinson's disease, aggregation is associated with amyotrophic lateral sclerosis (ALS), frontotemporal dementia (FTD), spinocerebellar ataxias, prion diseases, and several systemic amyloidoses. In ALS, proteins such as TDP-43, SOD1, and FUS can aggregate in motor neurons, disrupting RNA processing, protein homeostasis, and mitochondrial function. FTD often involves aggregates of tau, TDP-43, or FUS in the frontal and temporal lobes, leading to personality changes, impaired executive function, and language difficulties. Spinocerebellar ataxias, many caused by polyglutamine-expanded proteins, destabilize protein structure and promote aggregation in cerebellar neurons, resulting in impaired coordination, gait instability, and speech problems. Prion diseases, including Creutzfeldt-Jakob disease, occur when misfolded prion proteins induce normal prion proteins to adopt the abnormal β-sheet-rich form, leading to self-propagating aggregation and rapid neurodegeneration. Outside the nervous system, light-chain amyloidosis involves misfolded immunoglobulin light chains that deposit in organs, impairing their function.

In AA amyloidosis, the aggregated protein is serum amyloid A, usually associated with chronic inflammation. These deposits can impair organ function and lead to renal failure, cardiomyopathy, and fatigue.

Transthyretin amyloidosis (ATTR) happens when the transthyretin tetramer becomes destabilized, due to genetic mutations or age-related changes. Transthyretin, mostly made by the liver, carries thyroid hormone and vitamin A. When the tetramer breaks apart, misfolded subunits can form amyloid fibrils that build up in peripheral nerves and heart tissue, causing neuropathy and restrictive cardiomyopathy.

Dialysis-related amyloidosis results from the buildup of β2-microglobulin, a protein that is not efficiently cleared during long-term dialysis. It can deposit as amyloid in joints and bones, causing pain, stiffness, and carpal tunnel syndrome.

In the eye, cataracts form when crystallin proteins in the lens unfold and clump together. Age-related damage or genetic mutations destabilize these proteins, causing them to aggregate and scatter light, gradually impairing vision. Common features across these conditions include the destabilization of native protein structures, the formation of toxic aggregates or amyloid fibrils, failure of quality-control systems like chaperones and proteasomal degradation, and the selective vulnerability of specific tissues despite widespread protein expression.

Understanding protein folding, misfolding, and aggregation is crucial for uncovering the molecular causes of many diseases and for developing treatments. A major challenge is that aggregates often come from the body's own proteins, making them hard to target without disrupting normal functions. Nevertheless, research into aggregation pathways, chaperones, degradation systems, and toxic aggregate structures provides hope for therapies that can prevent or reverse these cellular problems. The cell's ongoing, vital effort to maintain protein order can sometimes fail; when it does, serious disease can occur.

Protein folding is essential to life's molecular machinery, but when it fails, the consequences can be profound. Misfolded proteins often expose hydrophobic surfaces normally buried within the protein core. These exposed regions promote clumping with other misfolded proteins, producing aggregates ranging from small, soluble oligomers to large, insoluble fibrils and plaques. These aggregates are not inert; they can disrupt cellular function, damage tissues, and contribute to cell death.

Huntington's disease is an inherited neurodegenerative disorder caused by a mutation in the huntingtin gene, which produces a huntingtin protein with an expanded glutamine repeat. This extended polyglutamine tract makes the protein prone to misfolding and aggregation. Abnormal huntingtin accumulates in the nucleus and cytoplasm of neurons, especially in the striatum and cerebral cortex. These aggregates are believed to disrupt gene transcription, protein degradation, and axonal transport, leading to progressive neuronal dysfunction and death. Huntington's disease is characterized by involuntary jerky movements, cognitive decline, and psychiatric symptoms that worsen over time.

The link between protein misfolding and cancer is gaining recognition. Many proteins involved in cell growth, proliferation, and DNA repair can become mutated or misregulated, leading to misfolding or aggregation. For example, the p53 tumor suppressor, often called the "guardian of the genome," is mutated in many human cancers. Some mutations cause p53 to misfold and form aggregates, decreasing its tumor-suppressing ability. Misfolded proteins can also promote tumor growth by interfering with apoptosis, changing signaling pathways, or supporting the formation of new blood vessels. Additionally, cellular stress and inflammation caused by widespread protein misfolding may create an environment that encourages cancer development and progression.

The connection between protein misfolding and disease highlights how sensitive biological systems are to molecular structure. A protein's three-dimensional shape determines what it does. When that shape changes, the protein might lose its normal function or gain harmful new properties, like aggregation and toxicity. The challenge for research and treatment is to understand how different proteins misfold, which cellular factors encourage or stop this process, and how misfolded proteins harm cells. These molecular stories show both what keeps us healthy and what causes problems in disease.

Intrinsically Disordered Proteins (IDPs)

Throughout much of the twentieth century, biology followed a simple rule: a protein must fold into a precise three-dimensional shape to function. This idea, often summarized as "structure determines function," influenced how scientists understood enzymes, receptors, and molecular machines. As experimental techniques improved in the late twentieth century, researchers discovered a growing class of proteins that challenged this rule. These molecules, called intrinsically disordered proteins (IDPs), lack stable structures under physiological conditions yet remain active and essential. Their discovery demonstrated that a well-defined fold is not always necessary for biological function.

Unlike folded proteins, IDPs exist as dynamic ensembles that constantly fluctuate among many conformations. This flexibility is not a flaw; it is essential to their function. Many IDPs become more ordered only upon binding to a specific target, a process known as folding upon binding. This behavior allows a single disordered protein to interact with multiple partners, often adopting different shapes for different ligands. Such adaptability makes IDPs especially effective in signaling, regulation, and control, where rapid responses and reversible interactions are critical. Their structural plasticity enables them to act as molecular hubs, integrating information from multiple pathways.

HERO proteins, short for Heat-Resistant Obscure proteins, are a recently discovered group of intrinsically disordered proteins first identified in 2020 by researchers at the University of Tokyo. They stay soluble and functional under extreme heat that would typically denature most folded proteins. Unlike globular proteins with stable three-dimensional structures, HERO proteins lack fixed secondary and tertiary structures in physiological conditions and instead exist as flexible, dynamic ensembles. Their amino acid composition is rich in charged residues like lysine, arginine, glutamate, and aspartate, and low in hydrophobic residues such as leucine, isoleucine, valine, and phenylalanine. This composition helps prevent hydrophobic collapse and aggregation, allowing HERO proteins to stay soluble even after prolonged exposure to temperatures above 95°C. Their extended, flexible structures provide large interaction surfaces, enabling them to act as molecular shields or chaperone-like protectors. Their lack of a fixed structure can also make them resistant to proteolytic degradation under certain conditions.

HERO proteins help protect cells by preventing other proteins from aggregating or denaturing under stress. Research suggests that they can shield enzymes and protein complexes from heat-induced damage and clumping. They seem to work through a molecular-shield mechanism, where long disordered regions interact with vulnerable proteins and stop aggregation-prone contacts. HERO proteins may also defend against desiccation and oxidative stress. Studies further indicate that they can decrease the toxicity of aggregation-prone proteins linked to neurodegenerative diseases such as amyotrophic lateral sclerosis (ALS) and Huntington's disease.

The properties of HERO proteins have garnered interest in biotechnology and medicine. Their ability to stabilize other proteins could help extend the shelf life and thermal stability of therapeutic proteins, industrial enzymes, and vaccine components that depend on cold-chain logistics. In neurodegenerative disease research, their capacity to reduce aggregation indicates a potential therapeutic approach. Researchers are also investigating their use in synthetic biology, where protein stability is often a limiting factor. As our understanding advances, HERO proteins may contribute to improving biopharmaceuticals, diagnostics, and resilient biological systems for industrial and medical applications.

The study of intrinsically disordered proteins has expanded our understanding of biology by illustrating that flexibility and disorder, alongside rigid structures, are fundamental principles of life. Protein folding has direct implications for disease: unfolded or misfolded proteins can clump together into toxic structures that interfere with cell function. Learning how to prevent, reverse, or control misfolding continues to be a major focus of biomedical research.

CHAPTER 5.

Proteins at Work: A Spectrum of Biological Functions

In this chapter, we explore several proteins with specialized functions in living systems, including some mentioned earlier. Although all proteins are made from the same amino acids, differences in their sequences and three-dimensional structures allow them to perform a wide range of functions. The examples below demonstrate the variety of protein roles: catalysis, structural support, transport, cellular communication, regulation, and mechanical force. By studying individual proteins and their functions, we can see how structure determines function and how proteins serve as precise, adaptable molecular machines within cells.

Hemoglobin in Action

Life relies on a steady supply of oxygen. This vital gas, inhaled with each breath, must travel from the lungs, where it is plentiful, to tissues, where it fuels cellular respiration. Hemoglobin, found inside red blood cells, performs this task with remarkable accuracy. It is not a passive carrier but a dynamic molecular machine designed to capture oxygen in the lungs and deliver it to tissues with lower oxygen levels. Hemoglobin is a tetrameric protein composed of four globin subunits arranged as two alpha (α) chains and two beta (β) chains in the main adult form, HbA ($\alpha_2\beta_2$). Each alpha chain contains 141 amino acids, and each beta chain contains 146. Each subunit is associated with a heme prosthetic group containing a central ferrous iron (Fe^{2+}) atom. Therefore, one hemoglobin molecule has four globin chains, four heme groups, and four iron atoms, enabling it to bind up to four oxygen molecules. The subunits are organized as two αβ dimers that coordinate to enable cooperative oxygen binding, allowing efficient oxygen loading in the lungs and release in tissues. Other physiological variants include fetal hemoglobin (HbF, $\alpha_2\gamma_2$), which has a higher oxygen affinity and is predominant before birth; the minor adult form HbA_2 ($\alpha_2\delta_2$); and embryonic hemoglobins containing zeta (ζ), epsilon (ε), or gamma (γ) chains during early development.

Hemoglobin's brilliance lies not only in its ability to carry oxygen but also in how it does so. Cooperative binding explains this efficiency. When hemoglobin encounters the high oxygen levels in the lungs, the first oxygen molecule binds to one heme group, triggering a small but significant change in the tetramer's shape. This shift alters the remaining heme sites, increasing their affinity for oxygen. Like a ferry at a busy dock, the first passenger makes it easier for the next passengers to board. As more sites become occupied, hemoglobin binds oxygen more readily. In the oxygen-rich lungs, this cooperative effect helps hemoglobin become nearly saturated, maximizing oxygen transport.

Once oxygenated hemoglobin enters the bloodstream, it travels to tissues where oxygen levels are lower because cells continually consume oxygen for metabolism. In this environment, the same cooperative mechanism promotes oxygen release. When one oxygen molecule leaves a heme group, hemoglobin changes shape and decreases the oxygen affinity of the remaining sites, making it easier for the other oxygen molecules to detach and diffuse into surrounding tissues. Like a ferry arriving at its destination, the first departing passenger makes it easier for others to leave. The lower the oxygen level in a tissue, the more oxygen hemoglobin releases, ensuring that cells receive what they need for vital functions.

This binding-and-release cycle is highly responsive to physiological conditions. One crucial factor is blood pH. In metabolically active tissues, carbon dioxide (CO_2) from cellular respiration dissolves in blood and forms carbonic acid, which dissociates into protons (H^+) and bicarbonate ions. The resulting drop in pH decreases hemoglobin's affinity for oxygen. This process, known as the Bohr effect, facilitates oxygen delivery to areas where it is most needed. In tissues that produce more CO_2, hemoglobin releases oxygen more easily. In the lungs, where CO_2 is expelled and pH is higher, hemoglobin's oxygen affinity increases, promoting efficient oxygen uptake.

Another key regulator is 2,3-bisphosphoglycerate (2,3-BPG), a byproduct of glycolysis present in red blood cells. 2,3-BPG binds mainly to deoxygenated hemoglobin and stabilizes that form, decreasing oxygen affinity. During extended exercise, high altitude exposure, or certain respiratory and heart conditions, red blood cells can raise their 2,3-BPG levels. This shifts the oxygen-hemoglobin dissociation curve to the right, encouraging oxygen release to tissues. Essentially, 2,3-BPG acts as a release facilitator, helping hemoglobin deliver oxygen when demand increases.

Hemoglobin's ability to carry oxygen is vital for survival and influences performance, adaptation, and health. At high altitude, where atmospheric oxygen pressure is lower, the body responds by producing more red blood cells, increasing hemoglobin levels, and enhancing oxygen transport. Changes in 2,3-BPG production also aid acclimatization by promoting oxygen release in tissues.

Hemoglobin's importance becomes clear when it malfunctions. Mutations in globin chains can lead to hemoglobinopathies. Hemoglobin also buffers blood pH by binding protons produced during CO_2 metabolism, reducing harmful drops in acidity. Additionally, hemoglobin can bind CO_2 at sites separate from oxygen-binding sites, forming carbaminohemoglobin and aiding in transporting CO_2 from tissues to the lungs for exhalation. Therefore, hemoglobin is vital not only for delivering oxygen but also for transporting carbon dioxide and maintaining acid-base balance.

From its tetrameric arrangement to its heme groups and iron atoms, every feature of hemoglobin is suited to its task. Without this remarkable protein, cells would be starved of oxygen, and complex life as we know it would be impossible.

The Defense System

The immune system is a marvel of biological engineering, constantly protecting against microscopic threats. Among its most precise and adaptable tools are antibodies. These Y-shaped molecular sentinels, introduced earlier, are central to the adaptive immune response, which learns from past encounters and remembers them. Unlike innate defenses, which offer broad, nonspecific protection, antibodies are highly specialized proteins that can identify and neutralize a wide range of targets with remarkable precision.

Antibodies are produced by specialized white blood cells called B lymphocytes. Their Y-shaped structure is perfectly suited to their function. Each antibody consists of two identical heavy chains and two identical light chains, as shown in Figure 12. These chains fold into specific regions that facilitate both target recognition and immune signaling.

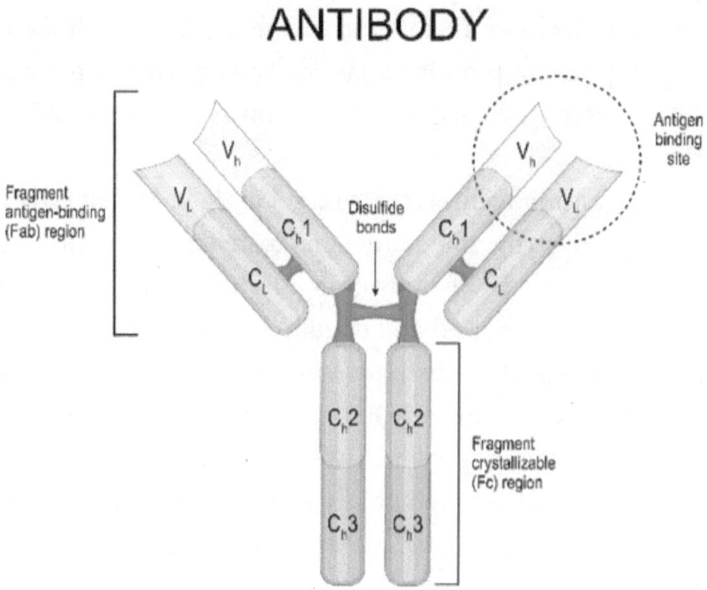

Source: Alamy, standard license, reference 0Y0111307071.

Figure 12. Structure of an immunoglobulin.

The tips of the Y-shaped arms, called the variable regions or Fab regions, are responsible for molecular recognition. These regions vary from one antibody to another and are shaped to bind specific molecular features on foreign invaders like bacteria, viruses, fungi, and toxins. The stem of the Y, known as the constant region or Fc region, is more consistent across antibody classes and helps signal other immune components to eliminate the targeted threat.

Antibody function depends on molecular recognition and specificity. The variable regions are shaped to match a particular molecular feature on a pathogen, known as an antigen. Antigens may be proteins, carbohydrates, or lipids that the body recognizes as foreign. For example, a protein on the surface of a flu virus may serve as an antigen, and an antibody with the right variable region can bind it precisely. This specificity allows the immune system to target harmful invaders while avoiding healthy cells.

Once an antibody binds to its target antigen, it generally does not directly destroy the pathogen. Instead, it functions as a molecular marker. In opsonization, antibody-coated pathogens are more easily identified and engulfed by phagocytic cells such as macrophages and neutrophils. Antibodies help these immune "clean-up crews" to find and eliminate invaders more effectively.

Antibodies also neutralize threats. They can bind toxins or viral surfaces, blocking their ability to interact with host cells. For example, if antibodies coat a virus, they can prevent it from attaching to and entering a cell. Similarly, antibodies can bind bacterial toxins, rendering them harmless.

Antibodies can also activate broader immune responses. When they bind to specific bacterial surfaces, they can trigger the complement system, a network of plasma proteins that can lyse bacteria or enhance inflammation, attracting more immune cells to the infection site. This coordinated response demonstrates how proteins work together in complex biological defense systems.

The variety of antibodies a person can produce reflects the complexity of the immune system's molecular machinery.

Although the basic Y-shaped structure stays the same, the amino acid sequences in the variable regions can vary greatly due to genetic recombination. This process generates billions of different antibodies, each with a unique antigen-binding site, equipping the immune system to recognize a vast array of foreign substances throughout life.

This diversity occurs during B-cell development through V(D)J recombination, a process where gene segments encoding the variable regions of heavy and light chains are randomly rearranged. The resulting variation alters key amino acids within the antigen-binding site, creating a vast repertoire of antibody specificities.

A major breakthrough in harnessing antibody specificity was achieved with monoclonal antibody technology, developed by César Milstein, Georges Köhler, and colleagues, whose work earned them the Nobel Prize in Physiology or Medicine in 1984. Their hybridoma technique fused antibody-producing B cells with immortal myeloma cells, creating hybrid cells that produce large quantities of identical antibodies targeting a single antigen. This innovation revolutionized diagnostics, therapeutics, and biomedical research. Later advances in humanization and recombinant antibody engineering significantly expanded clinical applications.

Antibody diversity is not only about recognizing many threats; it also enhances the quality of the response. When a B cell encounters its specific antigen, it becomes activated and multiplies. During this process, additional genetic changes can happen in antibody genes, improving binding strength. This process, called affinity maturation, produces antibodies that bind their targets more effectively as the immune response advances.

Antibody binding produces several important outcomes beyond opsonization and neutralization. In antibody-dependent cellular cytotoxicity (ADCC), an antibody binds to a target cell, such as a virus-infected or cancer cell, and immune cells, such as natural killer cells, recognize and kill the antibody-coated target. The antibody acts as a bridge between the target cell and the immune effector cell (Figure 13).

Different antibody classes, or isotypes, further enhance immune function. IgG, IgM, IgA, IgE, and IgD differ in their constant regions, distribution in the body, and effector functions. IgM is often produced early during infection and effectively activates complement. IgA is abundant in mucosal secretions such as saliva, tears, and breast milk, where it helps protect entry points. IgE is involved in allergic responses and defense against parasitic worms. Class switching enables B cells to customize antibody production to suit the needs of each immune response.

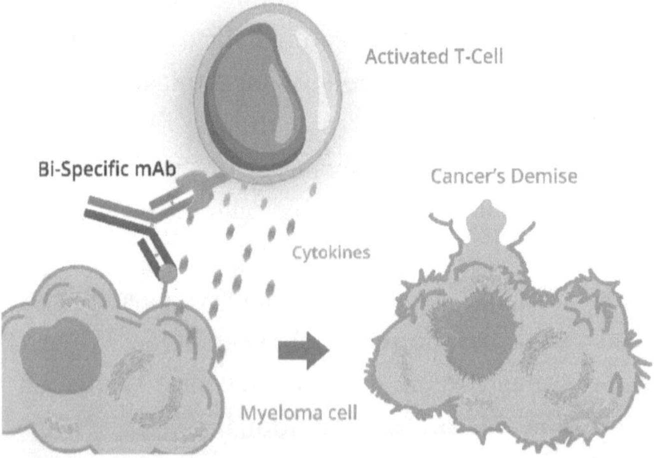

Source: Alamy under license OY0111526110.

Figure 13. Immunotherapy against cancer. A bi-specific monoclonal antibody binds to the immune cell and the cancer cell, leading to cancer cell demise.

In essence, antibodies are complex molecular tools that demonstrate the adaptive immune system's ability to learn, remember, and accurately target threats. Their Y-shaped structure provides highly variable antigen-binding regions for specificity, while their constant regions attract other immune components to coordinate defense. They are among the body's most effective security systems against unseen microbial threats.

PD-1: Cancer's Immune System Disguise

PD-1 (Programmed Death-1) is a protein that acts as a brake on the immune system, helping prevent immune cells from attacking healthy tissues. Cancer cells can exploit this safety mechanism to evade immune attack. When tumor cells engage the PD-1 pathway, they send a "do not attack" signal to T cells, which normally identify and destroy abnormal cells. This molecular disguise allows tumors to grow despite immune surveillance. PD-1 inhibitors, specialized therapeutic antibodies, block this pathway and help restore the immune system's ability to recognize and attack cancer cells. These therapies have revolutionized treatment for several cancers by unmasking tumor cells. However, cancer can still adapt. Some tumors produce soluble PD-1-related molecules that circulate in the bloodstream and may interfere with therapeutic antibodies. This molecular arms race continues to drive research in immuno-oncology, as scientists seek better ways to harness immune defenses and turn some cancers into manageable chronic diseases.

Force and Movement

From the flight of a bird to the motion of bacteria, the ability to move and generate force is a hallmark of life. In animals, much of this movement depends on two protein partners introduced earlier: actin and myosin. Their coordinated action drives the shortening and relaxation of muscle fibers, enabling movements from subtle gestures to powerful locomotion.

The way actin and myosin generate force is described by the sliding filament theory. Muscle contraction doesn't occur because individual filaments become shorter; instead, overlapping filaments slide past each other. Imagine two intertwined sets of ropes. Pulling one set through the other shortens the whole bundle without shortening any single rope. Actin and myosin perform a similar action, but through a cyclic process powered by molecular energy.

Actin filaments are long, double-helical strands composed of many actin subunits (Figure 14). These form the thin filaments in muscle cells. Among them are thick filaments mainly made of myosin. Each myosin molecule has a tail and a head. The tails join to form thick filaments, while the heads reach outward and interact with actin. These myosin heads function as molecular motors: they attach to actin and change shape to generate force. Their cyclic interaction is driven by ATP hydrolysis. Under resting conditions, regulatory proteins like troponin and tropomyosin block the myosin-binding sites on actin. When a muscle receives a contraction signal, these sites are exposed, allowing myosin heads to bind to actin and form cross-bridges.

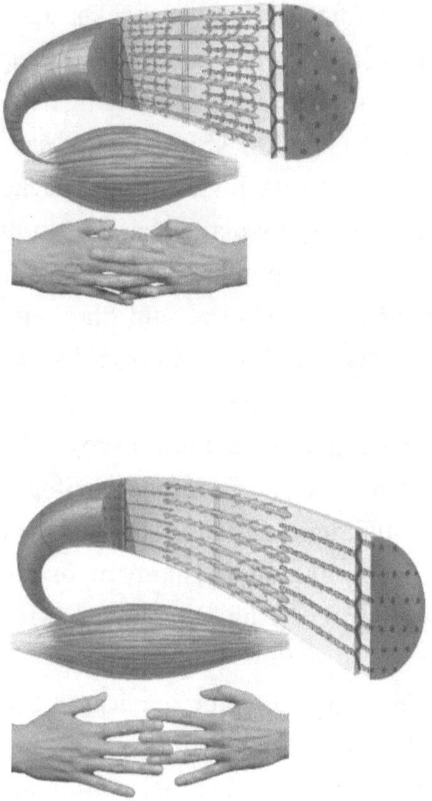

Source: Alamy under license: OY0116089081

Figure 14. Muscle cells relaxed and contracted. Hands show how fibers intertwine to cause contraction.

Myosin binding to actin causes the release of inorganic phosphate (Pi), followed by ADP, from the myosin head. This release triggers a major conformational change known as the power stroke, a bending movement that pulls the actin filament along the myosin filament. This is when force is generated, and filament sliding starts. After the power stroke, the myosin head remains attached to actin. Then, a new ATP molecule binds to myosin, weakening the interaction and allowing detachment. The ATP is hydrolyzed to ADP and Pi, which re-energizes the myosin head and puts it back into its high-energy "cocked" position, ready to attach to a new site on actin. If actin-binding sites stay exposed, the cycle of binding, power stroke, detachment, and re-cocking repeats. Each cycle pulls the thin filaments toward the center of the sarcomere, the basic contractile unit of muscle, causing the muscle fiber to shorten.

This process can be likened to a microscopic ratchet system or a coordinated tug-of-war. In the ratchet analogy, each power stroke moves the filament forward and prevents slipping backward. In the tug-of-war analogy, many myosin heads pull on actin filaments together, creating enough force to shorten the muscle. The coordinated action of thousands of actin-myosin cross-bridges within a muscle fiber generates the forces that move our limbs and support body weight.

The control of this powerful machinery is equally impressive. In skeletal muscle, contraction begins with nerve signals. When a nerve impulse reaches a muscle fiber, it causes the release of calcium ions (Ca^{2+}) from internal stores. Calcium then binds to troponin, changing its shape and moving tropomyosin away from the myosin-binding sites on actin. Once these sites are exposed, myosin heads can attach and start the contraction cycle. When the nerve signal stops, calcium is pumped back into storage, troponin and tropomyosin return to their resting positions, the binding sites are blocked again, and the muscle relaxes.

ATP is essential in this process. Without a continuous ATP supply, myosin heads would stay attached to actin in a contracted state, similar to rigor mortis after death when ATP production stops. This demonstrates the vital role of cellular respiration and ATP synthesis in muscle function. Different muscle types adapt the same actin-myosin mechanism for specialized roles. Cardiac muscle contracts rhythmically and continuously to pump the heart, requiring an efficient ATP supply. Smooth muscle, found in the walls of organs like the digestive tract and blood vessels, contracts more slowly and can maintain tension longer, using actin and myosin with distinct regulatory mechanisms.

The efficiency and power of the actin-myosin system are extraordinary. It enables movements ranging from the delicate precision of threading a needle to the strength required to lift heavy objects. Through coordinated protein interactions powered by ATP, chemical energy is converted into mechanical work. This partnership between actin and myosin is central to biomechanics and exemplifies proteins functioning as machines.

Enzymes and Their Catalytic Power

In the chemistry of life, countless reactions must occur quickly and accurately. Enzymes, a specialized group of proteins, make this possible. These biological catalysts speed up the biochemical processes necessary for living organisms, from food digestion to DNA replication. Without enzymes, many reactions would happen too slowly to keep cells alive. They sustain the flow of energy and matter that defines biological systems, transforming raw materials into the building blocks and energy sources required for cellular function.

An enzyme's catalytic power depends on its three-dimensional structure, especially a specific region called the active site. This site is not a random indentation but a carefully shaped pocket or groove designed to bind particular reactant molecules, called substrates. The chemical environment within the active site, created by specific amino acid residues, promotes the conversion of substrates into products. Enzymes accelerate reactions by lowering activation energy, the minimum energy required to begin a chemical reaction. One way to picture activation energy is as the effort required to push a boulder over a hill; enzymes lower the hill, making the reaction easier to start.

Enzymes lower this barrier by stabilizing the transition state, the unstable intermediate structure that briefly forms during a reaction. By binding to and stabilizing this state, enzymes increase the likelihood of product formation and greatly accelerate the reaction (Figure 15). As a result, biological chemistry can proceed under the mild temperature and pH conditions found inside cells.

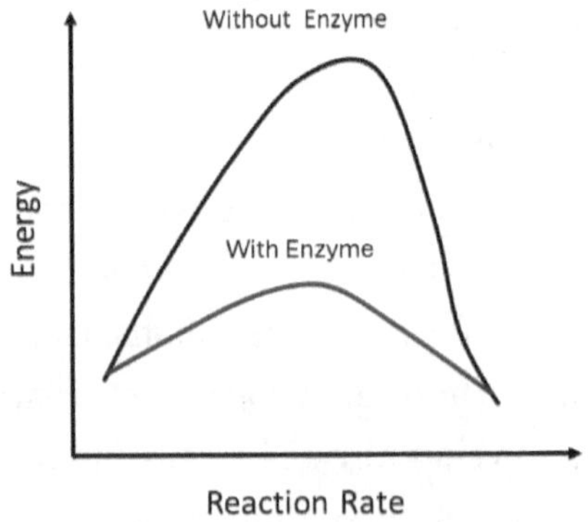

Figure 15. Comparison of a chemical reaction in the absence and the presence of an enzyme.

The relationship between an enzyme and its substrate is often explained through two complementary models. The classic "lock and key" model, proposed by Emil Fischer in the late nineteenth century, compares the enzyme's active site to a lock and the substrate to a key. Only a substrate with the proper shape can bind effectively. Later, Daniel Koshland's "induced fit" model provided a more dynamic view. In this model, the active site is flexible. When the substrate binds, the enzyme slightly changes shape, like a glove molding around a hand. This adjustment positions the substrate and catalytic amino acids more accurately, stabilizes the transition state, and enhances catalytic efficiency. Therefore, enzymes are not passive holders of substrates; they actively participate in the reaction they catalyze.

The diversity of life is reflected in the variety of enzymes, each designed for a specific task. Metabolism, which is the total of all chemical processes in an organism, relies on enzymes at every stage. Catabolic pathways break down complex molecules into simpler ones and release energy, while anabolic pathways construct complex molecules from simpler components and need energy. Enzymes make both types of pathways possible.

In carbohydrate metabolism, enzymes such as amylase begin digestion by breaking starch into smaller sugars. Amylase is found in saliva and pancreatic juice and helps break down dietary carbohydrates into energy sources. Deeper in metabolism, hexokinase catalyzes the first step of glycolysis by phosphorylating glucose. This traps glucose inside the cell and prepares it for energy production, a crucial step in nearly all organisms.

In protein metabolism, peptidases and proteases break down proteins into amino acids, which can then be used for new protein synthesis or energy generation. Conversely, aminoacyl-tRNA synthetases are vital for protein synthesis. Each enzyme attaches a specific amino acid to its corresponding transfer RNA (tRNA), helping ensure that the genetic code is accurately translated into a functional protein (Figure 7, Chapter 3). A single mistake in this process can produce a nonfunctional or harmful protein.

Lipid synthesis and breakdown also rely on enzymes. Lipases hydrolyze triglycerides into fatty acids and glycerol, which can be used for energy or stored as fats. Fatty acid synthase, a complex multi-enzyme system, constructs long hydrocarbon chains needed for membranes, signaling molecules, and energy storage.

Among the most important metabolic processes are the synthesis and use of nucleic acids, including DNA and RNA. Nucleases, like DNases and RNases, break down nucleic acids, while polymerases, such as DNA polymerase and RNA polymerase, generate new DNA and RNA molecules. These enzymes protect, copy, and express genetic information through replication, transcription, and repair. The accuracy of DNA polymerase is especially impressive; in many systems, proofreading and repair mechanisms lower errors to very low levels, helping to maintain genetic stability across generations of cells.

Topoisomerases are another crucial class of enzymes. They modify DNA topology by relieving supercoiling, untangling DNA, and relaxing strained DNA strands. To achieve this, they temporarily break the DNA backbone, allow controlled strand movement, and then reseal the break. Type I topoisomerases create single-strand breaks, while Type II topoisomerases make double-strand breaks and pass an entire DNA duplex through the opening. Both types utilize conserved catalytic domains and form transient covalent enzyme-DNA intermediates. Key tyrosine residues in their active sites form temporary phosphotyrosyl bonds with DNA, enabling strand passage while maintaining genetic integrity.

Topoisomerases are vital for DNA replication, transcription, recombination, and chromosome segregation. During replication, they relieve topological stress ahead of the replication forks, preventing excessive supercoiling that could stall polymerases. During transcription, they reduce positive supercoiling ahead of RNA polymerase and negative supercoiling behind it, enabling efficient gene expression. Type II topoisomerases are also essential during mitosis because they decatenate tangled DNA molecules and help resolve topological constraints during chromosome packaging and segregation.

Because topoisomerases are essential in rapidly dividing cells, they serve as important targets in cancer treatment. Topoisomerase I inhibitors, such as irinotecan and topotecan, trap the enzyme in a covalently bound state with DNA, leading to replication fork collisions and cell death. Topoisomerase II inhibitors, including etoposide and doxorubicin, stabilize enzyme-DNA cleavage complexes and can induce apoptosis. These drugs tend to be highly effective against cancer cells because such cells often divide quickly and rely heavily on topoisomerase activity. Although topoisomerases are highly conserved across living organisms, structural differences between bacterial and eukaryotic enzymes allow for selective targeting. For example, bacterial DNA gyrase, a Type II topoisomerase that introduces negative supercoils, is targeted by fluoroquinolone antibiotics. Ongoing research aims to develop more selective inhibitors, improve management of drug resistance, and gain deeper insights into topoisomerase roles in epigenetic regulation and chromatin remodeling.

Catalase, present in nearly all organisms exposed to oxygen, quickly converts hydrogen peroxide, a harmful byproduct of cellular respiration, into water and oxygen. This process prevents oxidative damage. Carbonic anhydrase, found in red blood cells and many tissues, speeds up the conversion of carbon dioxide and water into carbonic acid, a reaction vital for transporting carbon dioxide and maintaining pH balance.

Controlling enzyme activity is as important as catalysis itself. Cells regulate when and how strongly enzymes act, preventing wasteful reactions and coordinating metabolic pathways. Regulation can occur by controlling enzyme production, modulating activity with inhibitors or activators, or localizing enzymes to specific cellular compartments. Many enzymes are produced as inactive precursors, called zymogens or proenzymes, and are activated only when needed. Pepsinogen, the inactive precursor of pepsin, is activated by the acidic environment of the stomach, helping restrict digestive activity to the proper location.

Enzyme inhibitors can be reversible or irreversible. Reversible inhibitors bind noncovalently and can detach, allowing enzyme activity to resume. Competitive inhibitors bind at the active site and compete with the substrate, while noncompetitive inhibitors bind elsewhere, often at an allosteric site, and alter the enzyme's shape or function. Irreversible inhibitors form covalent bonds that permanently deactivate the enzyme. Many toxins and drugs work by inhibiting enzymes. For example, penicillin blocks a bacterial enzyme needed for cell wall synthesis, and several chemotherapy drugs inhibit enzymes essential for cancer cell growth.

Enzymes are not just facilitators; they are vital machinery that enable biological systems to harness energy, build structures, communicate, and react to their environment. Without their precise and efficient action, the biochemical processes that support life would slow down significantly or cease. Each enzyme, with its uniquely shaped active site and finely tuned catalytic machinery, provides an evolutionary solution to a specific chemical problem. Collectively, enzymes drive metabolism and keep life's processes dynamic.

Rubisco, or ribulose-1,5-bisphosphate carboxylase/oxygenase, is arguably one of the most important enzymes on Earth and the most abundant protein in the biosphere. Found in plants, algae, and photosynthetic bacteria, Rubisco catalyzes the first major step of carbon fixation in the Calvin cycle. It helps attach carbon dioxide to ribulose-1,5-bisphosphate (RuBP), producing two molecules of 3-phosphoglycerate that can be used to make glucose and other carbohydrates. This reaction is the gateway through which atmospheric carbon enters the biological world. By capturing enormous amounts of carbon dioxide each year, Rubisco links inorganic carbon to the food chains that support much of life on Earth. Yet despite its importance, Rubisco is surprisingly inefficient. It works slowly compared with many enzymes and has difficulty distinguishing between carbon dioxide (CO_2) and oxygen (O_2), which are similar in size and shape. When Rubisco binds oxygen instead of carbon dioxide, it triggers photorespiration, a wasteful process that reduces photosynthetic efficiency.

This limitation reflects Rubisco's evolutionary history. When the enzyme first evolved, Earth's atmosphere had abundant carbon dioxide but very little free oxygen, so there was little pressure for Rubisco to sharply distinguish between the two gases. As oxygen built up during events like the Great Oxygenation Event, Rubisco encountered oxygen more often and increasingly catalyzed the alternative reaction. Plants then had to recycle the resulting compounds through photorespiration, an energy-consuming pathway that lowers the overall efficiency of photosynthesis.

To compensate for Rubisco's slow rate and frequent errors, plants produce large amounts of the enzyme. In some leaves, Rubisco can make up a significant portion of total soluble protein, making it one of the most common enzymes on Earth.

While enzymes orchestrate the chemical reactions fundamental to life, protein function extends far beyond catalysis. Proteins are versatile cellular tools that perform tasks essential to survival, structure, communication, and regulation. To appreciate the proteome's full range, we must look beyond metabolic enzymes and examine other protein roles.

Structural Support

One of the most foundational roles of proteins is structural support. Just as a building requires a strong framework, cells and organisms depend on structural proteins to maintain shape, integrity, and resilience. These proteins often assemble into filaments or fibers that resist mechanical stress and support living tissues.

Collagen is the main structural component of connective tissues, forming strong and flexible fibers in skin, tendons, ligaments, cartilage, and bone. Its triple-helical structure, made of three polypeptide chains twisted together, provides excellent tensile strength. This strength allows tendons to connect muscles to bones and helps skin resist tearing. Collagen production requires post-translational modification, including hydroxylation of proline and lysine residues, followed by stabilizing interactions that maintain the triple helix. When this process is impaired, such as in vitamin C deficiency, collagen function is compromised, leading to disorders like scurvy and other connective tissue issues.

Keratin is another vital structural protein. Found in hair, nails, feathers, horns, and the outer layer of skin, keratin forms tough, fibrous structures with α-helical or β-sheet structures, depending on the organism and tissue. These structures provide durability, protection, and resistance to mechanical stress, explaining the toughness of hair and nails.

Within cells, the cytoskeleton provides a dynamic internal framework. Actin and tubulin form microfilaments and microtubules, respectively. Actin filaments help control cell shape, muscle contraction, and cell division, while microtubules serve as tracks for intracellular transport and form the mitotic spindle that separates chromosomes during cell division. These cytoskeletal elements are not static; they assemble and disassemble as needed, allowing cells to change shape, move, divide, and repair tissues.

Proteins are also indispensable for transport. They act as molecular couriers, moving essential molecules across cellular membranes and throughout the body. This is especially important because the lipid bilayer of the cell membrane is impermeable to many water-soluble substances.

Transport proteins embedded in membranes can be broadly classified as channels or carriers. Channel proteins form hydrophilic pores that allow specific ions or small molecules to pass rapidly. Ion channels are essential for nerve impulses and muscle contraction because they regulate the movement of sodium, potassium, calcium, and chloride ions across membranes. Aquaporins, another class of channels, facilitate rapid water movement and help maintain fluid balance in cells and tissues.

Carrier proteins bind specific molecules and change shape to move them across membranes. Transport may be passive, requiring no energy, or active, requiring ATP. A classic active transporter is the sodium-potassium pump (Na^+/K^+-ATPase), which moves three sodium ions out of the cell and two potassium ions into the cell for each ATP molecule hydrolyzed. This continuous pumping maintains steep Na^+ and K^+ gradients across the plasma membrane. These gradients regulate cell volume, drive secondary active transport, generate membrane potential, and support electrical signaling in nerve and muscle cells. The energy cost of this pump is substantial, underscoring its importance in cellular physiology.

At the organismal level, hemoglobin is a key transport protein, carrying oxygen from the lungs to tissues and helping return carbon dioxide to the lungs. Lipoproteins such as LDL and HDL transport cholesterol and triglycerides through the bloodstream, delivering energy-rich molecules and membrane building blocks while helping regulate lipid levels.

Proteins also serve as intermediaries in cellular communication, relaying information within and between cells. This signaling network coordinates processes ranging from embryonic development to nervous and endocrine system responses. Many hormones are proteins or peptides. Insulin, for example, helps regulate blood glucose by signaling cells to take up glucose, while growth hormone stimulates growth and cell reproduction. These hormones circulate through the bloodstream and bind specific receptor proteins on target cells, initiating intracellular signaling cascades.

Receptor proteins are central to signaling. Located on the cell surface or inside the cell, they bind specific ligands such as hormones, neurotransmitters, or growth factors. Ligand binding changes receptor shape and activates downstream pathways. G protein-coupled receptors (GPCRs), the largest family of cell-surface receptors, participate in vision, smell, neurotransmission, immune responses, and many drug responses. A GPCR is a single polypeptide that crosses the membrane seven times. When a ligand binds outside the cell, the receptor changes shape and activates an intracellular G protein. The activated G protein then interacts with effectors such as enzymes or ion channels, amplifying the signal and producing cellular responses such as altered gene expression, enzyme activity, or cell shape.

Within cells, proteins also mediate signal transduction. Kinases and phosphatases add or remove phosphate groups from other proteins, respectively, creating a reversible molecular switch that turns cellular processes on or off. Small GTPases such as Ras act as additional molecular switches, relaying signals from cell-surface receptors to internal pathways and, ultimately, to the nucleus.

A large group of proteins regulates gene expression and cellular behavior. These regulatory proteins serve as gatekeepers of the genome, determining which genes are active, when they are active, and how strongly they are expressed. Transcription factors are key examples. They bind specific DNA sequences near promoters or at distant enhancers to increase or decrease RNA production. The p53 protein, often called the "guardian of the genome," is activated by DNA damage. It turns on genes that pause the cell cycle, allowing repair. If damage is too severe, p53 can activate genes that trigger programmed cell death, preventing potentially cancerous cells from continuing to divide.

Regulatory proteins also control metabolism by influencing enzyme activity or gene expression. The bacterial lac operon is a classic example. This coordinated system contains genes needed for lactose uptake and breakdown, along with regulatory DNA sequences that control their expression. In the absence of lactose, the lac repressor binds the operator and blocks transcription. When lactose is available, allolactose binds the repressor, changes its shape, and reduces its affinity for the operator. RNA polymerase can then transcribe the operon, producing enzymes required to metabolize lactose. Additional regulation by glucose levels ensures that bacteria preferentially use glucose when it is available. Together, the lac operon shows how regulatory proteins and DNA elements help cells conserve energy and respond efficiently to changing conditions.

The Hidden Orchestra: Discovering Microproteins in "Junk" DNA

Only about 1% of the human genome encodes traditional proteins: the large, recognizable molecules that power cells. Much of the remaining genome was long dismissed as "junk DNA," evolutionary residue with no clear purpose. Hidden within this landscape, however, are small open reading frames (smORFs), short DNA sequences that can produce microproteins. These tiny proteins, often only a few dozen amino acids long, remained difficult to detect because standard tools such as mass spectrometry and genome-annotation software often miss molecules below certain size thresholds. A newer machine-learning tool called ShortStop is helping change that by identifying microproteins that may be biologically meaningful.

Developed at the Salk Institute, ShortStop acts as a sophisticated filter trained to distinguish likely functional microproteins from genetic noise. It compares real smORFs with computer-generated decoys, learning which sequences are most likely to encode functional products. Instead of testing thousands of candidates experimentally, researchers can prioritize the most promising targets. Applied to lung cancer data, ShortStop identified a previously unknown microprotein that was more active in tumors than in healthy tissue, suggesting possible value for early detection or targeted therapy.

Microproteins may be small, but they appear to play outsized biological roles. Recent studies link them to muscle development, cellular stress responses, and gene regulation. Many are conserved across species, suggesting evolutionary importance. Some act like molecular switches, binding larger proteins and changing their behavior; others fine-tune gene expression or serve as cellular signals. Their discovery challenges the assumption that biological importance depends on molecular size.

The medical implications are significant. Diseases such as Alzheimer's disease, diabetes, and cancer involve complex regulatory networks, and microproteins may help explain why treatments work differently across individuals or why symptoms vary. Because tools such as ShortStop can analyze standard RNA-sequencing data, many laboratories can now search for these hidden proteins in existing datasets. What was once dismissed as "junk DNA" is increasingly revealing itself as a layered source of regulatory information, with thousands of microproteins still likely to be discovered.

A common theme emerges from these diverse functions: proteins are remarkably versatile and adaptable. The same basic building blocks, amino acids, can be arranged into molecules with very different shapes, properties, and biological roles. These roles are not isolated; they are integrated into a dynamic network that supports nearly every aspect of life.

The Jellyfish's Secret

Deep in the cold waters of the Pacific Northwest lives an unassuming jellyfish, *Aequorea victoria*, that holds one of nature's most remarkable secrets. This translucent creature can produce light within its cells. When disturbed, it emits a green glow that ripples across its body like an underwater aurora. For decades, this display puzzled scientists, who wondered how a simple marine animal could create such vivid light without heat or electricity.

The mystery was solved when researchers discovered Green Fluorescent Protein, or GFP. This protein acts like a tiny biological light bulb, absorbing blue light and emitting a distinctive green glow. Its structure is elegantly simple: a barrel-shaped framework protects a light-emitting center called the chromophore.

The chromophore forms through a chemical process where three specific amino acids within GFP interact and modify one another. These reactions produce a natural fluorescent compound that glows when exposed to the right wavelength of light. What makes GFP especially remarkable is that it creates this fluorescent center on its own, without needing external cofactors.

After GFP folds into its protective barrel, amino acids at positions 65, 66, and 67 undergo oxidation and cyclization, creating a chromophore that absorbs blue light near 395 nanometers and emits green light near 509 nanometers. The surrounding barrel acts as a protective cage, shielding the chromophore while allowing light to pass through.

The discovery and isolation of GFP revolutionized biological research. Scientists realized they could attach GFP to other proteins and use it as a molecular flashlight to monitor activity inside living cells. Today, GFP and its colorful relatives light up laboratories worldwide, helping researchers explore processes from cancer progression to memory formation. What started as a curious glow in a jellyfish has become one of the most invaluable tools in modern biology, demonstrating how careful observation of nature can lead to groundbreaking scientific advances.

The Outer Dense Fibers of the Sperm Tail

The outer dense fibers (ODFs) are specialized protein structures that create a protective cage around the core of mammalian sperm tails. Made up of several major proteins, including ODF1, ODF2, ODF3, and ODF4, these fibers surround the nine outer doublet microtubules of the axoneme, which is the central motor structure of the tail. ODFs are unique to mammalian sperm and are tailored for the mechanical demands of reproduction. Instead of acting as rigid rods, they provide both support and flexibility, helping the tail withstand physical stress while allowing the bending movements needed for swimming.

The outer dense fibers work closely with the axoneme's motor machinery to generate sperm motility. As dynein proteins attached to microtubules create sliding forces, ODFs help transmit and shape those forces along the tail, producing coordinated bending waves that propel the sperm forward. Their elastic properties allow the tail to bend and recoil like a whip. Defects in ODF proteins can severely impair sperm motility and cause infertility. For example, mutations affecting ODF1 can disrupt the connection between the sperm head and tail, rendering sperm immobile and unable to fertilize an egg. These fibers are therefore essential components of the sperm propulsion system.

PKZILLA-1: Nature's Colossal Protein

Imagine a protein so enormous that it surpasses all others known to science. PKZILLA-1, discovered in the marine alga *Prymnesium parvum*, or golden algae, is currently considered the largest protein found in nature. It exceeds titin, the giant muscle protein once regarded as biology's heavyweight champion. PKZILLA-1 has a mass of about 4.7 megadaltons and contains 45,212 amino acids. It stretches roughly 1,250 nanometers, making it hundreds of times longer than many typical proteins. Scientists have aptly nicknamed it the "Mount Everest of proteins."

What makes PKZILLA-1 remarkable is not only its size but also its complexity and purpose. This massive protein contains 140 enzyme domains, specialized functional units that work together like an assembly line. These domains coordinate a long sequence of chemical reactions to produce prymnesin, a potent toxin used by the algae as a chemical weapon. PKZILLA-1, therefore, functions like a microscopic factory, coordinating many catalytic steps within a single enormous molecule.

The discovery of PKZILLA-1 has significant implications for science and the environment. Understanding how this protein is constructed and how it functions could guide the development of complex medicines, materials, and biotech processes. Simultaneously, PKZILLA-1 plays a role in toxin production during harmful algal blooms, which can lead to serious fish kills and harm aquatic ecosystems. By studying the genes that encode PKZILLA-1, scientists aim to improve the prediction and management of these environmental events. This enormous protein reminds us that, even at microscopic levels, nature keeps pushing the boundaries of biological complexity.

CHAPTER 6.

Protein Machines and Cellular Dynamics

Individual proteins rarely work alone. Many of the cell's most important processes depend on coordinated protein assemblies known as molecular machines. These multi-subunit complexes perform tasks that no single protein could accomplish on its own. Like the difference between a single tool and a fully equipped workshop, they transform simple biochemical reactions into sophisticated cellular operations.

One of the cell's key molecular machines is the replisome, the protein complex that precisely copies DNA before cell division. Instead of a single protein, the replisome is a dynamic assembly of dozens of proteins, each with a specialized function. It can be compared to a well-organized construction crew, with different teams preparing the site, laying the foundation, building the structure, and installing vital systems.

At the core of the replisome is DNA polymerase, the enzyme introduced in the previous chapter that synthesizes new DNA strands. However, DNA polymerase cannot work alone. It depends on partner proteins that prepare the DNA template and coordinate replication. Helicase unwinds the DNA double helix, exposing the two template strands. Because unwinding causes torsional stress ahead of the replication fork, topoisomerase alleviates this tension by temporarily cutting one or both DNA strands, allowing rotation, and then resealing the break. These enzymes work together to enable DNA to be copied efficiently without becoming overly twisted or tangled. Helicase powers unwinding by hydrolyzing ATP. Once the strands are separated, single-strand binding proteins (SSBs) attach to the exposed strands, preventing reannealing and secondary structures that could hinder replication. In doing so, they stabilize the template and keep it accessible for DNA synthesis.

DNA polymerase can only add nucleotides to the 3′ end of a growing strand, so DNA synthesis always happens in the 5′→3′ direction. Because the two strands of the DNA double helix run antiparallel, replication occurs differently on each template. The leading strand is synthesized continuously in the same direction as the advancing replication fork. The lagging strand is synthesized discontinuously as the fork opens, creating short segments called Okazaki fragments. Primase, an RNA polymerase, starts synthesis on the lagging strand by making short RNA primers that provide the free 3′-OH group needed for DNA polymerase to start. Once an Okazaki fragment is made, RNase H removes the RNA primer, another DNA polymerase replaces it with DNA, and DNA ligase seals the remaining nick by forming a phosphodiester bond, joining the fragments into a continuous strand.

The replisome functions as an integrated molecular assembly whose components stay coordinated as the replication fork advances. Regulatory proteins ensure that replication starts at the right time and place and proceeds with remarkable precision. Because accurate DNA replication is crucial for maintaining genetic information, defects in this machinery can cause mutations that impair cell function and organism health. Another key molecular machine is the ribosome, the cell's protein-making factory discussed in Chapter 3 (see Figure 7). Ribosomes quickly and accurately translate genetic information into proteins. Their precision depends on the coordinated actions of ribosomal RNA (rRNA) and ribosomal proteins, which ensure proper codon–anticodon pairing, catalyze peptide-bond formation, and coordinate movement along messenger RNA. The numerous interacting components and their precise spatial arrangement make the ribosome one of the most complex molecular assemblies in the cell.

Among the most remarkable molecular machines is ATP synthase, the enzyme responsible for producing most of the cell's ATP, its main energy currency. Embedded in the inner mitochondrial membrane of eukaryotic cells or the plasma membrane of bacteria, this rotary motor converts energy stored in a proton gradient into the chemical energy of ATP. ATP synthase consists of two functional units, F_0 and F_1. The membrane-embedded F_0 complex contains a proton channel through which protons flow down their electrochemical gradient. Instead of simply diffusing through the enzyme, proton flow drives rotation of a central shaft that includes the membrane-bound c-ring, a circular assembly of c subunits. As the c-ring rotates, it transmits mechanical motion to the F_1 complex, which extends into the mitochondrial matrix or bacterial cytoplasm. The F_1 complex contains the catalytic sites responsible for ATP synthesis. It features a hexameric ring of alternating α and β subunits surrounding the rotating central stalk. Rotation of the stalk causes each β subunit to cycle through three conformational states: open, loose, and tight. These structural changes enable ADP and inorganic phosphate to bind, catalyze ATP formation, and release the newly synthesized ATP. Through this elegant rotary mechanism, ATP synthase continually converts electrochemical energy into chemical energy, providing ATP for nearly every cellular process.

The replisome, ribosome, and ATP synthase represent only a small portion of the molecular machines functioning within cells.

Other key complexes include the proteasome, which breaks down damaged or misfolded proteins; the cytoskeleton, a flexible network of protein filaments that offers structural support, promotes movement, and transports cargo inside the cell; and molecular motors such as kinesin and dynein, which travel along cytoskeletal tracks to deliver organelles and vesicles to specific destinations. Despite their different roles, these machines share a common principle: cooperative assembly. Individual protein subunits, each with unique structural and functional features, specifically interact to form larger, integrated complexes. Their coördinated actions, often involving carefully timed conformational shifts and mechanical movements, drive biological processes that greatly surpass the capabilities of any single protein. Over billions of years, natural selection has optimized these assemblies into highly efficient molecular systems whose precision and dependability rival, and in some cases outperform, human-made machines.

Although often described simply as a lipid bilayer, the cell membrane is far more than a passive boundary. It is a dynamic interface that regulates the movement of substances into and out of the cell, preserving the internal environment essential for life. As discussed earlier, this function depends largely on proteins embedded within or spanning the membrane. These membrane proteins act as channels, transporters, receptors, and signaling molecules, allowing the cell to maintain homeostasis while sensing and responding to its surroundings. Without them, the cell could neither preserve its internal composition nor communicate effectively with its environment.

The membrane's selective permeability depends largely on these proteins. Their structures allow them to span the hydrophobic interior of the lipid bilayer while interacting with water-based environments on both sides. In doing so, they form channels that allow water-soluble molecules and ions to cross an otherwise impermeable barrier. The membrane is therefore not merely a container but a regulated gateway that controls material exchange between the cell and its surroundings. Channel proteins create specific pores through which selected ions or small molecules pass. Their selectivity depends on the size, charge, and chemical nature of the substances they transport. For instance, ion channels distinguish among sodium (Na^+), potassium (K^+), calcium (Ca^{2+}), and chloride (Cl^-) ions with impressive accuracy. This specificity arises from the amino acids lining the channel pore. Hydrophilic side chains create a favorable environment for water and ions, while hydrophobic residues on the protein exterior interact with the surrounding lipid membrane. Many channels also contain a selectivity filter, a narrow region that recognizes ions by size, charge, and hydration shell, allowing only appropriate ions to pass. Most ion channels are gated, opening and closing in response to specific stimuli. This regulation allows cells to control ion flow with exceptional precision, a capability essential for electrical signaling, muscle contraction, secretion, and many other physiological processes.

Voltage-gated channels respond to changes in membrane potential. When the electrical difference across the membrane reaches a threshold, voltage-sensing regions within the protein rearrange and open the channel. These channels are indispensable for generating and propagating action potentials in neurons and muscle cells.

Ligand-gated channels open when a specific signaling molecule binds to the protein. Neurotransmitters, for example, bind receptors that directly control ion channels, allowing ions to flow across the membrane and rapidly transmit signals between nerve cells.

Mechanically gated channels respond to physical forces such as stretching, pressure, or vibration. They allow cells to detect touch, hearing, and changes in mechanical stress.

Thermally gated channels respond to temperature changes and contribute to the perception of heat and cold.

Together, these gating mechanisms enable cells to regulate intracellular ion levels, generate electrical signals, and respond quickly to chemical, mechanical, electrical, and thermal stimuli. Not all membrane proteins function as channels. Transporter proteins bind specific molecules and change shape to move them across the membrane. Because transporters physically interact with their cargo, they usually move substances more slowly than channels but with equal high specificity. Transporters are classified by their transport direction and energy requirements. Uniporters move one type of solute across the membrane, usually down its concentration gradient. This process, called facilitated diffusion, does not need metabolic energy but depends on the transporter's ability to speed up the movement of molecules that cannot easily cross the lipid bilayer. A well-known example is the glucose transporter (GLUT) family. These proteins facilitate glucose uptake into cells, allowing this vital fuel to cross the plasma membrane. Different GLUT isoforms are present in various tissues and have different glucose affinities, helping each tissue get an appropriate supply based on its metabolic needs. Neurons, for example, have high-affinity transporters that continue to import glucose even when blood glucose levels are low, while other tissues use transporters suited to different physiological conditions. More complex transporters move two solutes at the same time. Symporters carry both substances in the same direction across the membrane, while antiporters move them in opposite directions. An important symporter is the sodium-glucose cotransporter (SGLT), found in the lining of the intestine and kidney tubules. SGLT uses the energy stored in the sodium (Na^+) gradient to bring glucose into the cell against its concentration gradient. As sodium ions enter the cell down their electrochemical gradient, glucose is transported along with them, allowing the cell to accumulate glucose even when its internal concentration is higher than outside. This process is called secondary active transport. Instead of using ATP directly, SGLT exploits the sodium gradient created by the sodium-potassium pump (Na^+/K^+-ATPase). By coupling glucose transport to sodium entry, the cell converts energy stored in one concentration gradient into moving another molecule against its gradient.

Antiporters move two solutes in opposite directions. An important example is the sodium–calcium exchanger (NCX), abundant in cardiac muscle and many other cell types. NCX exports calcium ions (Ca^{2+}) while importing sodium ions (Na^+), helping maintain the low intracellular calcium concentrations required for proper muscle contraction, signaling, and other cellular functions. Together, symporters and antiporters couple the movement of one molecule to another, allowing cells to transport nutrients, remove waste, and maintain essential ion gradients.

Some transport processes need a direct energy source because they move solutes against their electrochemical gradients. These proteins, called ATP-driven pumps or primary active transporters, get energy by hydrolyzing ATP. The most well-known example is the sodium-potassium pump (Na^+/K^+-ATPase), discussed in the previous chapter. Found in the plasma membrane of nearly all animal cells, this pump exports three sodium ions and imports two potassium ions for each ATP molecule hydrolyzed. In doing so, it maintains the ion gradients necessary for nerve impulses, muscle contraction, nutrient uptake, and cell-volume regulation.

Other ATP-driven pumps move protons (H^+). In eukaryotic cells, proton pumps acidify internal compartments such as lysosomes and endosomes, creating the low pH needed to break down cellular waste and recycle macromolecules. In the stomach, a specialized proton pump releases hydrochloric acid into the gastric lumen, aiding digestion. In bacteria, proton pumps generate the proton motive force that powers ATP production and nutrient transport. By moving ions against their concentration gradients, these pumps establish electrochemical differences that drive many cellular processes.

Membrane proteins do more than just transport molecules; they also enable cells to sense and respond to their environment. Among the most important signaling proteins are enzyme-linked receptors, which convert extracellular signals into intracellular responses. Most of these receptors have a single membrane-spanning region, an extracellular domain that binds signaling molecules, and an intracellular domain that either has enzymatic activity or interacts with an enzyme.

A major group of enzyme-linked receptors is the receptor tyrosine kinase (RTK) family. When a growth factor or hormone binds, it causes two receptor molecules to pair up, or dimerize, bringing their intracellular kinase domains close together. The kinases then phosphorylate each other on specific tyrosine residues, activating the receptors and creating docking sites for signaling proteins inside the cell. These phosphorylated sites recruit proteins that start multiple pathways. One example is activation of phospholipase C (PLC), an enzyme that splits a membrane phospholipid to produce two second messengers: inositol trisphosphate (IP_3) and diacylglycerol (DAG). IP_3 moves through the cytoplasm and binds to receptors on the endoplasmic reticulum, causing calcium to be released into the cytosol. Calcium then acts as a versatile signaling molecule by binding to regulatory proteins that control muscle contraction, secretion, metabolism, and gene expression. At the same time, DAG remains in the membrane, where it activates protein kinase C (PKC). When PKC is activated, it phosphorylates many target proteins, including transcription factors that regulate gene expression. Once activated, these transcription factors enter the nucleus, bind to specific DNA sequences, and change genes involved in cell growth, differentiation, survival, and metabolism. Therefore, a single signaling molecule from outside the cell can ultimately lead to lasting changes in how a cell behaves.

Although simplified here, RTK signaling demonstrates a core principle of cell biology: external signals are converted into intracellular responses through a series of protein interactions and enzymatic modifications. Since these pathways control cell growth, specialization, and survival, they need to be strictly regulated. Continuous activation of RTKs or their downstream signaling proteins is linked to many cancers and other diseases.

The plasma membrane itself is equally dynamic. According to the fluid-mosaic model, membrane proteins are embedded in a fluid lipid bilayer, allowing many to diffuse laterally and interact with neighboring proteins. This mobility is essential for receptor dimerization, signaling complex assembly, membrane trafficking, and cell adhesion. Not all membrane proteins move freely, however. Many are anchored to the cytoskeleton or connected to the extracellular matrix, restricting movement and organizing them into specialized membrane domains. These domains concentrate particular proteins and lipids, increasing the efficiency and specificity of cellular signaling and transport. Ultimately, the membrane's selective permeability depends on coordinated protein activity. While the lipid bilayer allows small nonpolar molecules such as oxygen and carbon dioxide to diffuse across readily, it is a strong barrier to ions and most polar molecules. Channels, transporters, and pumps provide regulated pathways that allow these substances to cross while maintaining the cell's internal composition.

Without membrane proteins, cells could not maintain ion gradients, import nutrients, or eliminate waste. Homeostasis would quickly break down. Instead of just passive openings, membrane proteins constantly adjust their activity based on signals from inside and outside the cell, helping cells adapt while keeping a stable internal environment. Research into membrane proteins continues to reveal how complex cellular regulation is. Their structures and functions have provided key insights into physiology and have become major targets for therapy, highlighting their essential role in both normal biology and human disease.

Cells constantly communicate with their environment and with each other. This exchange of information controls nearly all aspects of cellular life, from growth and differentiation to metabolism, movement, and programmed cell death. Proteins are central to this communication system, functioning as receptors, signaling molecules, enzymes, and transcriptional regulators that receive, transmit, and interpret information.

As explained earlier, signaling typically starts when an external molecule binds to a receptor on the cell surface. The activated receptor then initiates an intracellular signaling cascade: a series of protein interactions where one signaling protein activates the next. Instead of happening all at once, the signal moves through multiple proteins, allowing it to be amplified, combined with other signals, and carefully controlled before reaching its target. Many signaling proteins are enzymes that modify other proteins by adding or removing phosphate groups. These reversible changes act as molecular switches, quickly changing a protein's activity, location, or interactions. Because each activated enzyme can modify many targets, a single receptor activation can lead to a significant intracellular response.

Most signaling pathways ultimately influence transcription factors, which are specialized proteins that bind specific DNA sequences and control gene expression. Once activated, these factors enter the nucleus, where they either promote or inhibit transcription. This process allows external signals to produce lasting changes in cell behavior by changing which proteins the cell produces. Embryonic development, tissue repair, immune responses, and cell differentiation all rely on this precise regulation of gene expression. Signal transduction can be likened to a relay race: the receptor detects the initial signal, intracellular proteins pass and amplify it, and transcription factors deliver the final instructions to the genome. Because each step requires exact timing and coordination, any disruption can distort the entire response. Signal amplification is a key feature of these networks: one activated receptor can influence thousands or even millions of downstream molecules. For instance, a single activated protein kinase can phosphorylate many target proteins, each of which may trigger additional enzymes or pathways. As a result, even weak external signals can lead to strong physiological responses.

Cell signaling pathways also communicate through crosstalk. Instead of functioning separately, they form highly interconnected networks where one pathway can enhance, suppress, or modify another. This integration allows cells to process multiple environmental signals at the same time. Signals such as nutrient levels, growth factors, mechanical forces, and interactions with neighboring cells are combined into a coordinated response tailored to the cell's current state. The accuracy and flexibility of these protein-based communication systems are essential for maintaining balance within the body.

When signaling pathways become too active, too inactive, or improperly coordinated, it can lead to developmental problems, metabolic diseases, immune issues, and cancer.

Maintaining Protein Stability in a Dynamic Environment

The inside of a cell is far from static. It is a busy, constantly changing environment where proteins are being created, modified, transported, and broken down while facing shifting temperatures, pH levels, and reactive chemicals. Despite these challenges, proteins must maintain their precisely folded 3D structures to function properly.

A protein's function directly depends on its structure. As mentioned earlier, its amino acid sequence determines how it folds into a unique three-dimensional shape, stabilized by many noncovalent interactions and, sometimes, disulfide bonds. This folded form creates active sites, binding pockets, and interaction surfaces needed for catalysis, molecular recognition, and the assembly of larger protein complexes. Even small disruptions in structure can impair a protein's function. Higher temperatures increase molecular motion, weakening the noncovalent interactions that hold the protein together and potentially causing it to unfold or denature. This explains why prolonged high fever stresses cells and why maintaining stable internal temperatures is essential. Changes in pH also cause issues. Altering hydrogen ion levels changes how amino acid side chains ionize, disrupting ionic bonds and hydrogen bonds that help maintain the protein's structure. Conditions like acidosis or alkalosis can impair enzyme activity, protein interactions, and many other cellular processes. Proteins are also vulnerable to damage from reactive oxygen species (ROS), chemically reactive molecules produced during normal metabolism. Although ROS play important signaling roles at low levels, excess amounts cause oxidative stress, damaging proteins, lipids, and DNA. ROS can oxidize sensitive amino acids like methionine and cysteine. Oxidation can alter protein shape, decrease enzyme activity, disrupt protein-protein interactions, and cause proteins to clump together. Environmental factors such as ultraviolet radiation, pollution, cigarette smoke, heavy metals, and certain pesticides can increase oxidative stress by producing more reactive species or overwhelming the cell's antioxidant defenses. Over time, oxidative damage contributes to aging and many diseases, especially neurodegenerative disorders. Therefore, maintaining protein integrity requires both protection from damage and effective repair mechanisms. Cells achieve this through several complementary methods. Firstly, many proteins are naturally stable because evolution has favored amino acid sequences that fold into low-energy, stable shapes under normal conditions. Proteins from heat-loving organisms exemplify this principle well. Living in environments like hot springs, these organisms produce proteins with particularly strong stabilizing interactions and tightly packed hydrophobic cores, allowing them to remain functional at high temperatures that would quickly denature most other proteins.

Protein stability is further supported by molecular chaperones, introduced in Chapter 3. Instead of determining a protein's final structure, chaperones help newly made or partially unfolded proteins reach or recover their correct shape while preventing harmful aggregation. Since some damage is unavoidable, cells also have advanced repair and quality-control systems. Certain enzymes can reverse specific types of oxidative damage. For example, methionine sulfoxide reductases (MsRs) convert oxidized methionine residues back into methionine, restoring the function of many damaged proteins.

Proteins that cannot be repaired are removed via the ubiquitin-proteasome system (UPS). In this process, defective or outdated proteins are marked with multiple ubiquitin molecules, a small regulatory protein that acts as a molecular tag for destruction. The tagged proteins are then recognized by the proteasome, a large complex of multiple subunits that unfolds and breaks them down into short peptides. These peptides are further broken into amino acids that can be reused for making new proteins. This ongoing surveillance system maintains proteostasis, which is the balance between protein production, folding, repair, and breakdown. By removing damaged proteins before they accumulate, the UPS helps sustain cellular function and prevents the buildup of potentially harmful protein clumps.

Protein Turnover and Cellular Homeostasis

Proteins are not permanent cellular components. Their lifespans vary widely depending on their functions and the cell's needs. Regulatory proteins involved in processes such as cell-cycle control or signal transduction are often short-lived, allowing their concentrations to change rapidly in response to cellular conditions. In contrast, structural proteins generally persist for much longer.

The controlled breakdown of proteins is just as important as their synthesis. By selectively degrading old, damaged, or regulatory proteins, the ubiquitin-proteasome system allows cells to continuously adjust their protein makeup. For example, quickly removing signaling proteins can shut down a signaling pathway just as effectively as activating it. This ongoing renewal helps cells respond rapidly to developmental signals, environmental changes, and metabolic demands.

The balance between protein synthesis, folding, repair, and degradation demonstrates the remarkable resilience of living systems. Proteins are not static structures but dynamic molecules whose activities are constantly monitored and adjusted. Their stability depends not only on their molecular architecture but also on an intricate network of quality-control systems that maintain the integrity of the proteome. Together, sturdy protein structures, molecular chaperones, repair enzymes, and degradation pathways ensure that functional proteins are available while damaged ones are efficiently removed.

When Protein Dynamics Fail

The precise coordination of protein folding, assembly, signaling, transport, and degradation is essential for cellular health. Disruptions at any point in these processes can compromise homeostasis and lead to disease.

Cellular signaling provides a clear example. Normally, extracellular ligands activate receptors only when appropriate, triggering tightly controlled signaling cascades. Mutations that make signaling proteins permanently active—remaining "on" even without an external signal—can cause uncontrolled cell growth, a key feature of many cancers. Mutations in the RAS gene family illustrate this idea. Since RAS proteins usually transmit signals from growth factor receptors, mutations that keep them constantly active continuously stimulate pathways that promote cell growth and division.

The opposite problem can also occur. Mutations that inactivate key signaling proteins may prevent cells from responding appropriately to essential developmental or immune signals, contributing to developmental abnormalities and certain immunodeficiency disorders.

Large protein assemblies are similarly vulnerable. Because their function depends on the correct assembly of multiple subunits, mutations affecting even a single component can impair the entire complex. Some inherited retinal diseases, for example, result from defects in protein complexes required for the continuous renewal of photoreceptor outer segments.

Proteostasis and Proteinopathies

The processes that maintain protein balance are called proteostasis. This includes protein creation, folding, transport, quality checks, and breakdown. When these systems fail or become overwhelmed, damaged and misfolded proteins build up, leading to a group of disorders called proteinopathies. Neurodegenerative diseases are well-known examples. In these diseases, proteins that should be refolded or broken down instead clump into insoluble deposits that disrupt cell function. In Alzheimer's disease, abnormal buildup of amyloid-β peptides and hyperphosphorylated tau protein disrupt neuronal communication and eventually cause neuron death. Parkinson's disease involves the loss of dopaminergic neurons in the substantia nigra and the formation of intracellular protein clumps called Lewy bodies, mainly made of α-synuclein. In Huntington's disease, a CAG repeat expansion in the HTT gene creates a huntingtin protein with a long polyglutamine stretch. This altered protein tends to misfold and cluster, especially in striatal neurons, leading to ongoing neurodegeneration. Although these diseases differ in their molecular details, they share a common core: the failure of cellular systems that maintain proteostasis. As damaged proteins build up, they harm normal cell functions and eventually cause cell death.

Dynamic Organelles and Protein Trafficking

Protein dynamics go beyond individual molecules to entire organelles. Mitochondria, the main sites of ATP production, are highly dynamic structures that constantly undergo fusion and fission. These processes help maintain mitochondrial function, distribute mitochondria throughout the cell, and remove damaged organelles. Although mitochondria have their own genome and protein-making machinery, most mitochondrial proteins are encoded by nuclear genes, made in the cytoplasm, and imported into the organelle via specialized protein complexes. Problems with protein import, folding, or degradation impair mitochondrial function, decrease ATP production, and play a role in aging, metabolic disorders, and many neurodegenerative diseases.

The cytoskeleton is equally dynamic. Composed mainly of actin filaments, microtubules, and intermediate filaments, it offers structural support while facilitating intracellular transport, cell division, and movement. These protein filaments are constantly assembled and disassembled based on cellular needs. Mutations impacting cytoskeletal proteins or their regulatory factors can disrupt these dynamic processes. Some forms of muscular dystrophy result from defects in proteins that stabilize muscle cells, while altered cytoskeletal dynamics contribute to cancer metastasis by allowing tumor cells to migrate and invade nearby tissues. Cells must also ensure that proteins reach the right destinations. Newly formed proteins contain molecular "address labels," known as signal sequences, which direct them to specific organelles such as the nucleus, endoplasmic reticulum, Golgi apparatus, mitochondria, lysosomes, or plasma membrane. Specialized receptor proteins recognize these signals and guide proteins along the correct transport pathways. Failures in intracellular trafficking can have severe consequences. If enzymes meant for lysosomes do not reach their target, for instance, waste products build up because they cannot be broken down. Such defects are characteristic of several lysosomal storage diseases, highlighting the importance of proper protein localization.

Dynamic Regulation of Gene Expression

Gene expression itself depends on the coordinated activity of numerous protein complexes. Transcription factors, chromatin-remodeling complexes, RNA polymerases, and many accessory proteins work together to regulate which genes are expressed at any given time. Mutations affecting these regulatory proteins or the signaling pathways that control them can profoundly alter cellular behavior, contributing to developmental disorders, metabolic diseases, and cancer.

The cell is not just a static collection of molecules but a dynamic, highly organized system maintained by proteins in constant motion. Proteins fold into exact three-dimensional structures, assemble into molecular machines, transport substances across membranes, transmit signals, catalyze chemical reactions, generate mechanical force, and control gene expression. They are continually produced, modified, transported, repaired, and broken down, enabling the cell to respond quickly to changing conditions while keeping internal stability. The impressive coordination of these processes shows billions of years of evolutionary development. Protein machines rarely act alone; instead, they operate as interconnected networks whose activities are coordinated throughout the cell. This flexible organization influences every part of cellular life, from DNA copying and energy production to communication, movement, and adaptation. When these systems work properly, they keep the fragile balance needed for life. When they malfunction, the effects can range from metabolic problems and developmental issues to cancer and neurodegenerative diseases. Understanding how proteins assemble, act, and work together not only deepens our knowledge of cell organization but also provides a foundation for modern medicine. As research continues to reveal the complexity of these molecular systems, it creates new chances to diagnose disease, develop targeted treatments, and better grasp the core mechanisms that support life.

CHAPTER 7.

Proteins in Medicine and Biotechnology: The Future is Molecular

As we have seen, proteins, the workhorses of the cell, are not only fundamental to life's processes but have also become essential tools in modern medicine. Using nature's healing agents, such as therapeutic proteins, marks a major shift in how we treat a wide range of diseases. This isn't just about copying biological processes; it's about directly utilizing well-understood proteins (often produced through genetic engineering and advanced biotechnological methods) to restore function, slow or halt disease progression, or even eliminate harmful agents. The exceptional specificity and power of proteins make them highly suitable for targeted therapies, offering a level of precision once thought impossible.

One of the most well-known and widely recognized therapeutic proteins is insulin. In people with Type 1 Diabetes, the body produces little to no insulin, a hormone essential for regulating blood glucose levels. Without insulin, glucose cannot enter cells for energy, leading to hyperglycemia and serious complications.

After its first successful clinical use in 1922, insulin therapy evolved steadily, but a major breakthrough came in the late 20th century with the introduction of recombinant human insulin.

In 1982, Humulin became the first genetically engineered human insulin to receive regulatory approval. It was produced using genetically modified bacteria, usually E. coli, engineered to express the human insulin gene, replacing older animal-derived insulins. This innovation reduced allergic reactions, improved purity, and enabled large-scale, consistent production.

Today, most insulin used worldwide is recombinant and falls into several categories:

1. Rapid-acting insulins (e.g., Insulin lispro, Insulin aspart): taken before meals to control blood sugar spikes.
2. Short-acting insulin (regular human insulin): used around mealtimes, though less common now.
3. long-acting insulins (e.g., Insulin glargine, Insulin detemir): provide a steady "basal" insulin level over 24 hours or longer.
4. Ultra-long-acting insulin (e.g., Insulin degludec): offers even more stable, flexible dosing.

These "analogs" are slightly modified forms of human insulin designed to act faster or last longer, improving glucose control and reducing the risk of hypoglycemia.

Modern insulin therapy is designed for patient self-management:

Insulin pens: prefilled, portable devices that simplify dosing and improve accuracy compared to traditional vials and syringes.

Syringes and vials: still widely used, especially where cost is a concern.

Insulin pumps: small wearable devices that continuously deliver insulin (basal rate) and allow on-demand doses for meals.

Closed-loop systems ("artificial pancreas"): combine a pump with a continuous glucose monitor (CGM) and automated dosing algorithms to adjust insulin in real time.

Recombinant insulin and modern delivery methods have transformed diabetes care from emergency treatments to precise, personalized management. Patients can now modify dosages based on meals, activity, and glucose levels, significantly improving quality of life and long-term health. This biotechnological advance enables patients to inject accurate insulin doses, effectively replacing their body's insufficient production. The journey from understanding insulin's vital role in glucose metabolism to its large-scale therapeutic production demonstrates the power of molecular biology and protein engineering. Being able to synthesize a complex protein like insulin with the same structure and function as the natural version has turned a life-threatening condition into a manageable chronic disease for millions worldwide. Current research continues to develop advanced insulin formulations, including long-acting and rapid-acting types, further tailoring treatment to individual needs and lifestyles. As a result, therapeutic proteins have become powerful tools in combating many diseases.

The immune system naturally produces antibodies, Y-shaped proteins that recognize and neutralize foreign invaders, including pathogens, as discussed in previous chapters. However, cancer cells can sometimes evade immune detection. These antibodies function like highly targeted guided missiles. Once attached to a cancer cell, they can activate various anti-cancer mechanisms: they can directly block signals that promote cancer cell growth, mark the cancer cell for destruction by other immune cells, or deliver toxic payloads to the cancer cell via antibody-drug conjugates.

Immunotherapy for cancer marks a major advancement in the ongoing fight against this complex group of diseases. Traditional treatments like chemotherapy, while often effective in reducing tumor size, operate in a largely non-specific way. Chemotherapy drugs target rapidly dividing cells, including not only cancer cells but also many healthy cells in the body, such as those in hair follicles, bone marrow, and the digestive tract. As a consequence, patients frequently face significant side effects like hair loss, fatigue, nausea, and a weakened immune system, making them more susceptible to infections. In many ways, chemotherapy can feel like trying to eliminate a single bird with a machine gun: powerful but imprecise and causing extensive collateral damage.

In contrast, immunotherapy utilizes the body's own immune system to recognize and attack cancer cells more precisely. One of the major breakthroughs in this field has been the development of monoclonal antibodies, lab-engineered molecules that are administered to patients intravenously (IV) or subcutaneously (SC) and bind to specific proteins on cancer cells. These treatments operate with much higher accuracy than traditional approaches, targeting malignant cells while sparing most normal tissues. A notable example is rituximab, a pioneering monoclonal antibody used to treat certain lymphomas and leukemias. Rituximab specifically targets the CD20 antigen, a protein present on the surface of B cells, including many cancerous ones. By binding to CD20, it effectively "flags" these cells, enabling the immune system to recognize and eliminate them. This targeted approach allows effective treatment while minimizing damage to other cell types.

Another key example is trastuzumab (Herceptin), used to treat HER2-positive breast cancers. These cancers overexpress the HER2 receptor, a protein that promotes cell growth and division. Trastuzumab binds to this receptor, blocking its signaling pathways and thereby slowing or stopping tumor growth. In addition, it can also recruit immune cells to attack the cancer, further enhancing its therapeutic effect.

The development of monoclonal antibodies and other forms of immunotherapy has opened new avenues for treating cancers and other diseases that were once very difficult to manage. These therapies exemplify the shift toward personalized medicine, in which treatments are tailored to a patient's tumor's specific molecular characteristics. As research continues, this approach holds great promise for improving survival rates and quality of life for cancer patients worldwide.

The impact of recombinant DNA technology, however, extends far beyond cancer treatment. One of its earliest and most influential medical applications was the production of recombinant human growth hormone (hGH) for children with growth hormone deficiency. Before recombinant hGH became available, the hormone was extracted from the pituitary glands of deceased individuals, a process that was expensive, limited in supply, and carried the risk of transmitting prion diseases. Recombinant production provided a safe, reliable, and virtually unlimited source of hGH, demonstrating how genetic engineering could transform the treatment of a wide range of human diseases.

The application of therapeutic proteins extends to treating Genetic disorders are often treated by replacing a missing or defective enzyme. Many inherited diseases result from mutations that lead to the production of nonfunctional enzymes. Without a working enzyme, specific metabolic pathways are disrupted, leading to toxic build-up or shortages of essential molecules. Enzyme replacement therapy (ERT) works by giving patients recombinant versions of the missing enzyme.

An example of this is Gaucher disease, a rare genetic disorder caused by a deficiency in the enzyme glucocerebrosidase (GCase). GCase is responsible for breaking down a fatty substance called glucocerebroside. When GCase is deficient, glucocerebroside builds up in various organs, especially the spleen, liver, and bone marrow, leading to symptoms such as organ enlargement, bone abnormalities, and anemia. Imiglucerase, a recombinant form of GCase, is given intravenously to patients, effectively replacing the missing enzyme and helping reduce the buildup of glucocerebroside. Similarly, treatments for other lysosomal storage diseases, such as Fabry disease, Pompe disease, and Niemann-Pick disease, developed by Genzyme Corporation, rely on enzyme replacement therapy with recombinant enzymes. Although these therapies are often lifelong and expensive, they can greatly reduce symptoms and improve quality of life for those affected by these serious genetic conditions. The development and production of these therapeutic proteins are closely connected to our understanding of protein structure and function. Likewise, the binding sites on antibodies that interact with antigens and the receptor-binding regions of protein hormones depend heavily on their specific protein shapes. Biotechnology has provided tools not only to produce these proteins but also to engineer them for greater stability, increased effectiveness, and reduced immunogenicity (the tendency to trigger an immune response).

From Gene to Medicine

The process of creating a therapeutic protein usually begins with identifying the target, such as a hormone deficiency, a cancer cell marker, or a missing enzyme. Next, the gene encoding the desired therapeutic protein is isolated or synthesized. This gene is then inserted into a host organism, such as bacteria, yeast, mammalian cells, or transgenic plants and animals, enabling the host to produce the protein in large quantities. This process, known as recombinant DNA technology, allows scientists to insert a gene encoding a desired protein into a host organism. These host cells function as tiny factories, transcribing and translating the foreign gene to produce large amounts of the target protein. This technique forms the foundation of biopharmaceutical manufacturing. After production, the therapeutic protein must undergo thorough purification to remove contaminants and ensure safety and effectiveness. Quality control measures are essential at every stage, from cell culture to the final product formulation (Figure 16).

The specificity of protein-based therapies is a key advantage. Unlike small-molecule drugs, which are simple, low-weight chemical compounds that can easily enter cells and circulate widely in the body, therapeutic proteins are much larger and more complex. Because of their small size (low molecular weight), these drugs can sometimes interact with multiple targets, leading to off-target side effects.

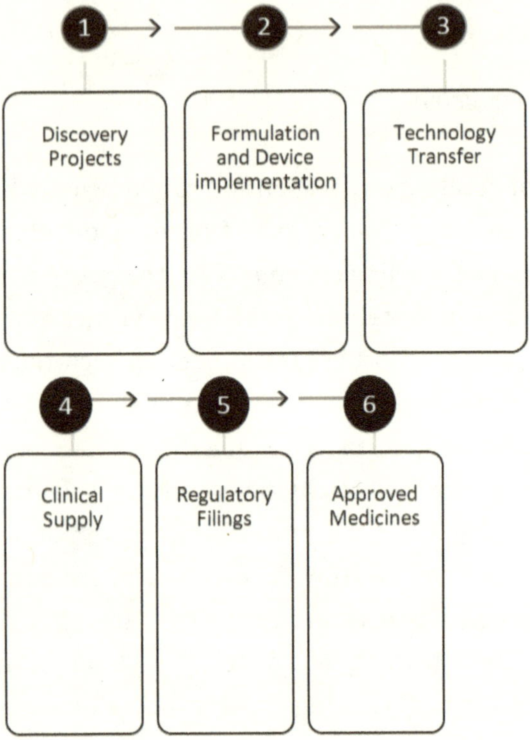

Figure 16. From discovery projects to approved medicines.

To put the size difference into perspective: a typical therapeutic protein, such as a monoclonal antibody, has a molecular mass of about 150,000 daltons, whereas a common small-molecule drug like Aspirin has a molecular mass of only about 180 daltons. That means a small molecule is roughly 800 times smaller. Many antibiotics, such as Penicillin, also fall within this low-molecular-weight range.

Because therapeutic proteins are large and structurally complex, they can be engineered to bind very specifically to a single biological target, much like a key fitting into a specific lock. This reduces unwanted interactions elsewhere in the body. Such precision minimizes the risk of side effects and enables more personalized treatment strategies. For example, antibody fragments offer some targeting advantages of full-length antibodies but are smaller, which may allow better tissue penetration and unique pharmacokinetic profiles. Bispecific antibodies, capable of binding two different targets at once, are being developed to engage immune and cancer cells, thereby boosting anti-tumor immunity. Proteins also serve as versatile carriers for drug delivery, providing a biocompatible, highly customizable alternative to synthetic carriers. Since proteins are naturally occurring biomolecules, they are generally well tolerated by the body and can be engineered with precise structural and chemical features. This makes them ideal "scaffolds" for encapsulating, protecting, and delivering therapeutic agents, such as small-molecule drugs, nucleic acids, and even other proteins, to specific tissues or cells. An especially promising example is silk-based biomaterials. Derived from silkworms or produced recombinantly, silk proteins (like fibroin) have outstanding mechanical strength, stability, and biodegradability. These qualities allow silk to be processed into various formats, including nanoparticles, hydrogels, films, and fibers. In drug delivery, silk can encapsulate therapeutic compounds and release them gradually over time. Its structure can be engineered to respond to environmental triggers, such as pH changes or temperature shifts, enabling targeted release at disease sites like tumors. Additionally, silk can be functionalized with specific ligands to improve targeting of certain cell types, further increasing its precision.

Collagen is also commonly used as a scaffold for drug delivery. As a key component of the extracellular matrix, collagen is naturally biocompatible and aids in cell adhesion and tissue integration. It can be fashioned into sponges, gels, or injectable matrices that serve as reservoirs for therapeutic agents. These collagen-based platforms are especially useful in regenerative medicine and localized drug delivery, such as delivering BMP-2 protein for bone formation, wound healing, or cancer treatment, where sustained release at a specific site is beneficial. By changing collagen's structure or crosslinking density, researchers can control how quickly a drug is released.

Beyond silk and collagen, progress in protein engineering has allowed the creation of completely new protein scaffolds. These engineered proteins can be designed to self-assemble into nanostructures that can carry drugs or to contain targeting regions that recognize specific receptors on diseased cells. For instance, fusion proteins can join a targeting component with a therapeutic payload, ensuring the drug is delivered exactly where it is needed while reducing systemic exposure.

Overall, protein-based drug delivery systems are a powerful combination of materials science, molecular biology, and medicine. Their biodegradability, specificity, and adaptability make them very appealing for next-generation therapies, especially in cancer treatment, where targeted and controlled delivery can greatly enhance effectiveness while lowering side effects.

The development of therapeutic proteins encounters many challenges. Production can be complicated and costly, requiring specialized facilities and strict regulatory oversight. Proteins are also susceptible to degradation in the body, which can limit their effectiveness and might require frequent dosing. Additionally, the human immune system can sometimes identify therapeutic proteins as foreign, triggering an immune response that can decrease their effectiveness or cause side effects. Still, ongoing research into protein engineering and formulation methods continuously tackles these issues, resulting in more stable, effective, and better-tolerated protein-based treatments.

The development of recombinant vaccines is another field where proteins have transformed public health. Traditional vaccines often use weakened or inactivated forms of viruses or bacteria. However, some of these vaccines still pose risks or are difficult to produce. Recombinant DNA technology allows for the production of specific viral or bacterial antigens—the proteins on a pathogen's surface that trigger an immune response—in a safe, controlled manner. These purified antigens can then be used as vaccines. For example, the hepatitis B vaccine is a highly effective recombinant protein vaccine. Instead of using the entire hepatitis B virus, the vaccine contains the virus's surface protein (HBsAg), produced in yeast cells. When injected, this protein prompts a protective immune response without any risk of infection. Similarly, parts of HPV (human papillomavirus) vaccines and some components of COVID-19 vaccines are made using recombinant protein technology. This approach provides significant advantages in vaccine safety, purity, and production scalability.

In conclusion, therapeutic proteins are a powerful and adaptable class of drugs that have significantly impacted medicine. From treating chronic conditions like diabetes with insulin to fighting aggressive cancers with monoclonal antibodies and correcting metabolic issues in genetic disorders with enzymes, these biomolecules are advancing personalized and targeted therapies. The ability to understand, engineer, and produce these complex natural remedies has transformed the treatment landscape, offering hope and an improved quality of life to millions of patients worldwide. As our understanding of protein science expands and biotechnological skills improve, we can expect even more innovative and life-saving protein-based therapies, highlighting their vital role in modern healthcare. The precision, potency, and specificity of protein molecules make them ideal for tackling some of the most complex and challenging diseases humanity faces.

Proteins as Diagnostic Tools

While we've examined the roles of proteins as therapeutic agents, these molecular workhorses also function as powerful diagnostic tools, providing subtle yet strong signals of our health status. The ability to detect and measure specific proteins, or their altered forms, in biological samples such as blood, urine, or tissue biopsies has transformed disease diagnosis, allowing for earlier detection and more targeted treatments than ever before. This diagnostic power comes from the unique structural and functional properties of proteins, which are highly responsive to physiological and pathological changes in the body.

Proteins, with their intricate three-dimensional structures, are highly specific in how they interact. This specificity is exactly what makes them valuable biomarkers. Just like a lock is made to accept only a specific key, many proteins in our bodies have unique binding partners, such as other molecules, cells, or small chemical signals. When disease occurs, these sensitive molecular interactions can be disrupted, new proteins can be produced, or the levels of existing proteins can change significantly. These changes form unique molecular signatures that can be detected using various biochemical and immunological tests.

One of the most important uses of protein biomarkers is in screening and early cancer detection. Abnormal proteins or excess proteins normally found at low levels can signal early signs of malignancy. A prime example is Prostate-Specific Antigen (PSA). PSA is a protein produced by cells in the prostate gland. In healthy men, PSA levels in the blood are usually low. However, in men with prostate cancer, benign prostatic hyperplasia (enlarged prostate), or prostatitis (prostate inflammation), PSA levels can rise significantly. Although PSA is not exclusively linked to cancer, other conditions can also raise its levels. A consistently high or rapidly increasing PSA level warrants further testing, such as a biopsy, to detect prostate cancer early and improve treatment outcomes. Advances in highly sensitive immunoassays have made PSA testing a routine part of prostate cancer screening for men over a certain age. The ongoing challenge and active area of research is to improve PSA as a biomarker, possibly by analyzing different PSA forms or combining PSA tests with other markers to improve diagnostic accuracy and reduce unnecessary biopsies.

Protein biomarkers are crucial for diagnosing acute conditions, especially in emergency situations. A typical example is diagnosing myocardial infarction, commonly called a heart attack. During a heart attack, heart muscle cells are damaged and release their intracellular contents into the bloodstream. Among these contents are specific proteins, mainly troponins. Cardiac troponins (especially troponin I and troponin T) are proteins essential for muscle contraction. When the heart muscle is injured, troponin leaks into the blood. Detecting elevated troponin levels has become the standard method for diagnosing a heart attack. The main benefit of troponin is its high specificity for cardiac muscle injury; it is not significantly elevated in other causes of chest pain. Additionally, troponin levels increase quickly after symptoms start and remain elevated for several days, allowing diagnosis even if a patient arrives hours after symptoms began. Modern diagnostic tools can deliver troponin test results within minutes, enabling swift treatment decisions crucial for saving heart tissue and improving patient outcomes.

The diagnostic value of proteins covers a wide range of diseases. For instance, in liver conditions, markers such as alanine aminotransferase (ALT) and aspartate aminotransferase (AST) are routinely tested. These enzymes are primarily found in liver cells, and their release into the bloodstream indicates liver cell damage. Elevated levels can signal various problems, including hepatitis, cirrhosis, or fatty liver disease. Similarly, bilirubin, a breakdown product of heme processed by the liver, can be measured. Increased bilirubin levels, which often cause jaundice (yellowing of the skin and eyes), may point to impaired liver function or bile duct blockage.

In autoimmune diseases like rheumatoid arthritis, specific autoantibodies can act as diagnostic markers. Autoantibodies are antibodies that mistakenly target the body's own tissues. For example, rheumatoid factor (RF) and anti-cyclic citrullinated peptide (anti-CCP) antibodies are often found in the blood of people with rheumatoid arthritis, aiding in confirming the diagnosis and evaluating disease activity.

Detecting infectious agents often relies on protein-based tests. While it's common to directly identify the pathogen's genetic material (DNA or RNA), many diagnostics focus on detecting specific proteins produced by the pathogen. For example, rapid antigen tests for influenza or strep throat detect viral or bacterial proteins respectively. Similarly, HIV tests often detect antibodies produced by the body in response to the virus, as well as specific viral proteins like p24. The accuracy and reliability of these protein-based tests have significantly improved over time, allowing for quick and dependable infection detection.

The power of protein biomarkers lies not only in detecting their presence or absence but also in measuring and analyzing their levels and modifications. For example, C-reactive protein (CRP) is a liver-produced protein generated in response to inflammation. Elevated CRP levels serve as a general marker of systemic inflammation and are linked to infections, autoimmune diseases, and injury. High-sensitivity CRP (hs-CRP) tests can identify even low levels of CRP, offering a more sensitive assessment of cardiovascular risk.

Another key aspect is identifying abnormal protein structures or isoforms. For example, in Alzheimer's disease, the abnormal accumulation of misfolded proteins like beta-amyloid and tau in the brain is a hallmark of the condition. While diagnosis mainly relies on clinical evaluation and imaging, ongoing research aims to detect specific forms of these proteins in cerebrospinal fluid (CSF) or blood as potential diagnostic and prognostic biomarkers. Recognizing these protein variants is a vital focus in biomarker research.

The development and use of protein biomarkers are closely connected to advances in biotechnology and analytical chemistry. Techniques such as enzyme-linked immunosorbent assay (ELISA), western blotting, and, more recently, mass spectrometry-based proteomics have played vital roles in identifying, measuring, and characterizing protein biomarkers. For example, ELISA is a popular method that uses antibodies to detect and quantify specific proteins in a sample. It works by attaching an antibody to a surface to capture its target protein from a fluid sample. A second antibody, often linked to an enzyme, is then used to detect the captured protein, producing a measurable signal—such as a color change—that correlates with the amount of protein.

Mass spectrometry (MS) offers a powerful, label-free method for discovering and validating biomarkers. It enables the simultaneous identification and measurement of thousands of proteins in complex biological samples. By analyzing the mass-to-charge ratio of ions, MS can identify proteins based on their unique molecular weights and fragmentation patterns. This high-throughput technique is essential for finding new biomarkers and confirming existing ones in large patient populations. Proteomics, the large-scale study of proteins, is rapidly advancing our understanding of the proteome (the complete set of proteins expressed by an organism or system) and how it changes in disease states.

The field of "liquid biopsy" is rapidly expanding and depends heavily on protein biomarkers. Instead of invasive tissue biopsies, researchers are investigating the detection of circulating tumor DNA (ctDNA), circulating tumor cells (CTCs), and, importantly, circulating proteins shed by tumors into the bloodstream. These circulating proteins can provide a less invasive method to diagnose cancer, monitor treatment response, and detect recurrence. For example, specific proteins released by tumors may be present in the blood at very low levels, but with increasingly sensitive detection technologies, they are becoming practical diagnostic targets.

The journey of a protein from a research discovery to a clinically validated biomarker is a rigorous process. It involves multiple stages:

1. *Discovery:* Identifying potential protein candidates through high-throughput screening of patient samples.
2. *Verification:* Confirming the presence and altered levels of these proteins in larger, independent patient cohorts.
3. *Validation:* Establishing the diagnostic accuracy (sensitivity, specificity, predictive values) of the biomarker in a clinical setting.
4. *Clinical Utility:* Demonstrating that the use of the biomarker leads to improved patient outcomes, such as earlier diagnosis, better treatment selection, or enhanced monitoring.

Challenges still exist in the field of protein biomarkers. Biological samples can be complex, and protein levels can be affected by factors unrelated to disease, such as diet, exercise, and other medical conditions. This can cause false positives or false negatives. Additionally, the low abundance of some key biomarkers in biological fluids can make detection technically difficult. Achieving enough sensitivity (the ability to correctly identify those with the disease) and specificity (the ability to correctly identify those without the disease) is essential for a reliable diagnostic test.

Despite these challenges, the future of protein biomarkers in medicine appears very promising. As our understanding of disease mechanisms advances and biotechnological tools become more sophisticated, we will identify and confirm more proteins as diagnostic and prognostic markers. This will lead to earlier, more accurate diagnoses, personalized treatments, and ultimately, improved health outcomes for patients. Proteins, with their natural specificity and ability to respond to physiological changes, are truly essential molecular sentinels, guiding us toward a future of proactive and precise healthcare.

The significant role of proteins in medicine is undeniable, shaping how we diagnose and treat diseases. However, their influence extends far beyond healthcare, sparking a transformative era in biotechnology. This revolution is driven by our growing ability to understand, isolate, and engineer proteins, unlocking their potential as vital tools across many industrial and commercial sectors. From the food we eat to the fuels that power our world, proteins are quietly but fundamentally transforming modern industry.

Proteins: Beyond Medicines

One of the most important roles of proteins in biotechnology is the incredible versatility of enzymes. These biological catalysts, known for speeding up specific biochemical reactions, have been widely used in industrial processes for many years. Their exact specificity allows them to carry out desired chemical changes without producing unwanted byproducts, which is a major benefit in many manufacturing settings. In the food industry, enzymes are essential tools that improve product quality, increase yield, and enhance processing efficiency. For example, amylases are crucial in baking because they break down starches in flour into simpler sugars that yeast can ferment, helping to improve bread texture and crust formation. In the dairy sector, rennet (a complex mixture of enzymes) is used to coagulate milk proteins, forming curds that are key to cheese making. Lactase is another essential enzyme, often added to milk to break down lactose, the sugar that many people find hard to digest, making dairy products easier to enjoy. Beyond these common applications, enzymes are also used in fruit juice clarification, brewing for better fermentation, and meat tenderization, as discussed in Chapter 1.

The impact of enzymes also extends to biofuels. As the world looks for sustainable energy sources, enzymes have become vital in converting biomass into biofuels like ethanol. Cellulases, a type of enzyme, are essential for breaking down the complex structure of plant cell walls into fermentable sugars. This step is crucial in producing cellulosic ethanol, a second-generation biofuel that uses non-food biomass such as agricultural waste and wood chips. The efficiency of these enzymatic reactions directly influences the economic viability of biofuel production, prompting ongoing research to find and develop more robust and cost-effective cellulases.

The manipulation of genetic material, a key aspect of modern biotechnology, has also transformed our ability to produce specific proteins for therapeutic and industrial uses. This breakthrough has had a major impact on medicine, allowing the production of human therapeutic proteins that were previously hard or impossible to obtain in sufficient amounts.

The ability to produce specific proteins in host systems has also enabled large-scale production of industrial enzymes.

Enzymes such as proteases (which break down proteins), lipases (which break down fats), and amylases (which break down starches) are now manufactured in enormous quantities for use in detergents, where they help break down stains; in animal feed to improve nutrient absorption; and in various chemical synthesis processes. Engineering these enzymes, often through directed evolution or site-directed mutagenesis, allows scientists to customize their properties, such as increasing stability under harsh industrial conditions (e.g., high temperatures or extreme pH) or boosting catalytic efficiency for specific substrates.

The economic impact of these advancements is substantial. The biotechnology sector, driven by protein engineering and recombinant production, is a multi-billion-dollar industry. It has not only introduced new therapeutic options and diagnostic tools but also spurred innovation in areas like agriculture (e.g., enzyme-based feed additives) and materials science (e.g., biodegradable plastics made with engineered enzymes). The ability to utilize proteins—once confined to living organisms—has transformed them into advanced industrial agents, promoting progress and innovation across multiple industries. The cost of developing a recombinant biological medicine from initial research through regulatory approval and market launch varies greatly, depending on the product, indication, clinical trial size, and other factors. Published estimates typically range from hundreds of millions to several billion dollars. Industry-sponsored analyses report average costs of about $1–3 billion, considering high failure rates of drug candidates and the cost of capital, while some studies suggest that total investments for particularly complex biologics can exceed $4 billion. These costs include discovery research, preclinical testing, large-scale manufacturing development, clinical trials (Phases I–III), regulatory approvals, and the setup of manufacturing facilities, which are especially expensive for biologics due to the need for advanced cell cultures, purification methods, and strict quality control.

The journey from understanding protein structure and function to engineering them for specific tasks represents a profound leap in our technological capabilities, underscoring the central role of proteins in shaping our future.

Ancient Protein Reveals Oldest Known Human Interbreeding Episode

For decades, scientists have primarily used ancient DNA to piece together the story of human evolution. However, DNA degrades relatively quickly, especially in warm environments, rendering many ancient fossils genetically silent. Proteins, on the other hand, can last much longer. In a groundbreaking new study, researchers in China examined proteins preserved in the dental enamel of six Homo erectus fossils dating approximately 400,000 years old. By extracting and analyzing ameloblastin, a protein crucial to tooth enamel formation, they identified two distinct variants. One of these, known as the M273V variant, was previously found only in Denisovans, the elusive human relatives closely related to Neanderthals. This shared molecular marker offered the first direct evidence that these ancient human groups once exchanged genes.

The discovery provides strong evidence that Homo erectus and Denisovans not only encountered each other in Eurasia but also successfully interbred. This marks the oldest known instance of interbreeding between human species, predating the later mixing of Neanderthals and modern humans by hundreds of thousands of years. Homo erectus was one of the most important species in human history: the first human ancestor to spread widely beyond Africa and one of the longest-lasting human species ever to exist. The finding shows that these ancient populations remained biologically compatible despite evolving along separate branches for long periods. Instead of existing as isolated species, ancient humans seem to have formed overlapping populations capable of producing fertile offspring when their paths crossed.

The implications for human evolution are profound. For much of the twentieth century, evolution was often pictured as a simple branching tree in which one human species replaced another in a neat sequence. Increasingly, however, evidence points to a far more interconnected history. The new protein data suggest that *Homo erectus* acted as an important genetic bridge linking multiple human lineages. Through a chain of interbreeding events, genes may have passed from *Homo erectus* into Denisovans, then from Denisovans into Neanderthals, and eventually into modern *Homo sapiens*. In this view, human evolution resembles a braided stream rather than a straight line, with populations repeatedly separating, reconnecting, and exchanging genetic material over vast stretches of time.

The study also emphasizes the increasing importance of proteins in uncovering aspects of human history that DNA alone cannot reveal. Since proteins are more durable than genetic material, they can preserve biological information from fossils that are much older than those still containing usable DNA. This opens the door to identifying additional "ghost populations" of ancient humans whose genetic traces only survive indirectly in modern populations. Some living groups in Asia and the Philippines already carry Denisovan ancestry, and this new research suggests that part of that inheritance may ultimately originate from Homo erectus. As methods for analyzing ancient proteins improve, scientists are likely to discover a more complex story of our origins—one where the human family tree is shaped by repeated encounters and exchanges spanning hundreds of thousands of years.

Reengineering Proteins

The ability to not only understand but also actively design and build proteins with specific, new functions is one of the most exciting areas in modern molecular science. This growing field, known as protein engineering, goes beyond just isolating or making naturally occurring proteins; it involves intentionally reshaping the fundamental components of life to perform tasks they were not originally designed for or to greatly improve existing ones. Think of it as moving from simply collecting rare biological artifacts to sculpting new tools from the very core of biological machinery. This hands-on approach to manipulating these molecular wonders is opening up a world of possibilities for medicine, industry, and basic research.

At its core, protein engineering depends on our increasing understanding of the complex link between a protein's three-dimensional shape and its biological role. By learning how amino acid sequences fold into specific forms and how these forms interact with other molecules, scientists can predict and make adjustments that alter a protein's behavior. This is akin to understanding the blueprint of a complex machine and then fine-tuning certain gears, levers, or sensors to produce a desired outcome.

One of the main goals of protein engineering is to enhance or modify enzyme catalytic activity. Enzymes are nature's most efficient catalysts, speeding up biochemical reactions with high precision. However, natural enzymes may not perform well under the harsh conditions of industrial processes, which often involve high temperatures, extreme pH levels, or inhibitors. Protein engineers can carefully alter an enzyme's amino acid sequence, especially at key sites within its active site or regions that influence overall stability. For example, they might add amino acid variants that bind a specific substrate more effectively, increasing the enzyme's efficiency. They might also modify the enzyme to resist unfolding (denaturation) at higher temperatures, extending its lifespan in industrial settings. This can involve replacing amino acids that break down at high heat with more robust ones or adding disulfide bonds to stabilize the structure, making it more durable. These modifications are made using recombinant DNA technology. The potential in the detergent industry is vast. Naturally occurring proteases, lipases, and amylases are already included in laundry detergents to break down protein stains (such as blood or grass), fat stains (such as grease or makeup), and starch stains (such as gravy or chocolate). Still, these enzymes must work effectively across a range of wash temperatures, from cold to hot, and in the presence of various surfactants and builders. Protein engineering has been crucial in developing enzymes that are not only highly active at lower temperatures (saving energy) but also stable across a wide pH range and resistant to breakdown by other detergent components. This leads to better cleaning, a smaller environmental impact, and lower manufacturing costs.

Beyond enhancing existing functions, protein engineering also involves designing entirely new ones. This is where the field truly becomes a form of custom molecular design. Scientists can create proteins that serve as biosensors, capable of detecting specific molecules such as disease biomarkers, environmental toxins, or chemicals, and producing a measurable signal in response. This often involves linking a recognition domain (which binds the target molecule) to a reporter domain (which produces a detectable output, like fluorescence or a color change). For example, a protein could be engineered to bind to a particular cancer cell surface marker. When binding occurs, a conformational change might trigger the release of a fluorescent molecule, enabling visualization and identification of cancer cells.

Antibodies, naturally produced by the immune system to target foreign invaders, can be engineered to have improved binding affinity to disease-causing agents or cancer cells. This targeted delivery approach is revolutionizing cancer therapy, for example, by creating "guided missiles" that seek out and destroy malignant cells.

The methods used in protein engineering are mainly divided into two categories: rational design and directed evolution.

Rational design is a top-down approach where scientists use their understanding of protein structure and function to make targeted, informed modifications. By using detailed structural data—obtained through techniques such as X-ray crystallography or Nuclear Magnetic Resonance spectroscopy—and computational modeling, they hypothesize which amino acid changes are most likely to produce the desired effect and then introduce those specific mutations. This process is similar to an architect drawing precise renovation plans, carefully choosing each new component. In contrast, directed evolution mimics natural selection but accelerates it in the lab. This method involves creating large libraries of protein variants, often by introducing random mutations into the gene encoding the protein. These variants are then tested or screened to find those with the desired trait. For example, if the goal is to develop an enzyme that performs better at high temperatures, the mutated enzyme library is incubated at high temperature, and only the enzymes that remain active—or show enhanced activity—are selected and further propagated. This cycle is repeated multiple times, gradually enriching the population with proteins that have improved features. It's like an accelerated artificial breeding program where only the strongest survive and reproduce. Directed evolution is especially useful when the specific molecular mechanisms behind a desired function are not fully known, as it leverages nature's trial-and-error process to achieve engineering goals.

Both rational design and directed evolution have their advantages and are often combined. Rational design provides precision and is effective for targeted changes, while directed evolution is better at exploring large sequence spaces to find new or improved functions, even when the specific rules for those improvements are initially unclear.

The impact of protein engineering is already clear across many industries. In medicine, it's not just about improving existing therapies but also about creating entirely new types of drugs and diagnostic tools. The ethical issues and challenges related to protein engineering are significant as well. As we gain the ability to design and produce new biological entities, we need to consider potential unintended consequences. This includes ensuring the safety of engineered proteins used in medicine and industry, preventing the accidental release of engineered organisms into the environment, and thinking about the broader societal effects of manipulating life at its molecular level. Careful testing, regulation, and open communication are essential to address these challenges responsibly.

Looking ahead, the future of protein engineering is one of increasing sophistication and expanding capabilities. Advances in computational protein design, artificial intelligence, and high-throughput screening continue to drive innovation. We are moving toward an era where proteins can be designed from scratch with specific properties, rather than just modifying existing natural proteins. This level of control over biological molecules promises to unlock solutions for some of humanity's most urgent challenges, from developing new cures for diseases to creating sustainable industrial processes and novel materials. Protein engineering is not just about understanding proteins; it is about harnessing their fundamental elegance and power to build a better future, molecule by molecule. It marks a major shift from observing and utilizing nature's creations to actively co-creating with it, pushing the boundaries of what is biologically and technologically possible. The potential applications are broad and diverse, making them hard to fully define; however, the core principle remains: by precisely modifying the building blocks of life, we can engineer solutions to many problems. This proactive, creative manipulation of protein architecture is undeniably shaping the future of medicine, biotechnology, and beyond, marking a pivotal moment in our relationship with the molecular world.

The horizon of protein science is a landscape painted with vibrant hues of innovation and potential, especially as we explore personalized medicine. For decades, medical treatments have mostly followed a one-size-fits-all approach, giving the same drugs and therapies to large groups of patients and assuming that biological responses are generally similar. However, we now recognize that this assumption is a major oversimplification. Each person has a unique biological fingerprint, a complex pattern formed by their genetic makeup, environmental exposures, and lifestyle. At the core of this individuality is the proteome, the full set of proteins expressed by an organism or cell at a specific time. Unlike the relatively stable genome, the proteome is constantly changing, responding to internal and external signals. It is this proteome that directly governs cellular function, disease development, and, importantly, an individual's reaction to medical treatments.

The growing field of proteomics, which studies proteins on a large scale, provides us with tools to explore these individual protein landscapes. Advanced mass spectrometry techniques, combined with sophisticated bioinformatics, enable researchers to identify and quantify thousands of proteins in biological samples like blood, tissue biopsies, and cellular extracts. By comparing proteomic profiles from healthy individuals to those of patients with specific diseases, or by analyzing how a patient's proteome responds to treatment, we can gain new insights into the molecular basis of health and disease. This detailed understanding helps lay the foundation for personalized medicine.

Imagine a situation where, before starting treatment for a complex disease like cancer or an autoimmune disorder, a patient undergoes a detailed proteomic analysis. This analysis would reveal not only the presence of disease-specific protein markers but also the levels of various enzymes, receptors, and signaling molecules that regulate cellular behavior. With this information, clinicians could select therapies most likely to be effective for that particular patient and, perhaps more importantly, avoid treatments predicted to cause severe side effects or be ineffective. For example, some cancer treatments target specific protein mutations or signaling pathways. A proteomic profile could determine if a patient's tumor expresses the target protein or relies on an alternative pathway, guiding the choice of the most appropriate targeted therapy. Similarly, in autoimmune diseases, understanding which immune proteins are dysregulated could lead to highly targeted immunotherapies that reduce abnormal immune responses without damaging the body's overall defenses.

Furthermore, proteomics can aid drug development and improvement. By understanding how different individuals process or respond to a drug at the protein level, pharmaceutical companies can develop more effective medications or predict which patient groups will benefit most from specific treatments. Two people with the same diagnosis may show very different underlying protein network patterns, influencing how they respond to treatments, lifestyle changes, or environmental exposures. Mapping these networks is a crucial step toward personalized medicine, where interventions are customized to an individual's unique biological profile rather than a one-size-fits-all approach. This has significant implications for clinical trial design, potentially making trials more efficient and successful by enrolling participants who are more likely to respond positively. The concept of a "biomarker" has long been central to diagnostics and prognostics, but proteomics is taking this idea to a new level of sophistication. We are moving beyond studying single biomarkers to a broader, more informative view: analyzing entire protein networks and their dynamic interactions within cells and tissues. Instead of focusing on one protein at a time, modern methods (often related to systems biology and proteomics) examine how hundreds or thousands of proteins communicate, regulate each other, and form interconnected pathways. This systems-level perspective is vital because most physiological states, whether health or disease, result not from a single molecular change, but from coordinated shifts across complex networks. Processes like inflammation, immune responses, metabolism, and cancer progression all depend on tightly controlled interactions among many proteins.

An important and often underappreciated part of this complexity is nutrition, especially the role of dietary proteins and their building blocks, amino acids. In health, a balanced and diverse intake of amino acids supports stable protein networks, efficient metabolism, and proper immune responses. In disease, however, these processes can become dysregulated. Cancer cells, for example, often rewire amino acid metabolism to support rapid growth, while chronic illness or malnutrition can interfere with protein synthesis and repair.

Adding another essential layer to this systems-level understanding is the microbiome, often called the "biome," especially the gut microbiome. This vast community of microorganisms, including bacteria, fungi, and viruses, plays a crucial role in shaping protein networks and overall physiology. The microbiome aids in digesting dietary components, such as proteins, and can convert amino acids into bioactive compounds that influence host metabolism, immune response, and even brain signaling. When healthy, a balanced microbiome supports efficient nutrient absorption, maintains immune stability, and helps preserve tissue health, such as the gut lining. It also interacts with host protein networks by producing metabolites that act as signaling molecules, linking diet, microbial activity, and human cellular functions. In disease, however, disruptions to this microbial community, known as dysbiosis, can cause widespread effects. Dysbiosis has been linked to conditions like inflammatory bowel disease, obesity, metabolic syndrome, and even cancer. These changes can alter amino acid processing, influence inflammation, and disrupt key protein signaling pathways. Importantly, diet is one of the strongest factors affecting the microbiome. Different protein sources and amino acid profiles can shape microbial composition and activity, which, in turn, feeds back into the host's biological systems. This creates a complex, multi-layered interaction between what we eat, the microbes inside us, and the protein networks that manage our physiology.

In summary, moving from single markers to the analysis of complex protein networks allows us to better capture the true intricacy of biological systems. When combined with insights into amino acid nutrition and the role of the microbiome, this systems-level perspective reveals a deeply interconnected picture of health and disease, one in which diet, microbial ecology, and molecular signaling all work together to shape individual outcomes.

The "symbiosis" of proteomics and protein engineering holds tremendous transformative potential. Proteomics uncovers the 'what' (the proteins and their quantities), while protein engineering concentrates on the 'how,' enabling the design and creation of proteins with specific functions based on proteomic insights. Beyond antibodies, engineered enzymes are set to significantly influence personalized medicine. For instance, if proteomic analysis detects an excess of a particular enzyme that contributes to disease, protein engineers can develop targeted inhibitors to precisely block its activity. Conversely, identifying deficiencies in vital enzymes allows engineered enzymes to be used as replacement therapies, restoring normal metabolic functions. Developing biosensors is another promising area where proteomics and protein engineering work together to improve personalized diagnostics. Imagine a small, wearable device that continuously monitors a patient's proteome in real-time, sensing subtle changes indicative of disease onset or treatment deviations. Such a device would likely use engineered proteins that bind selectively to disease biomarkers, triggering signals like fluorescence or electrical responses, providing immediate alerts to patients and clinicians. This proactive, continuous monitoring could revolutionize disease management by enabling early intervention before symptoms worsen or permanent damage occurs. It shifts the focus from reactive treatment to proactive prevention and personalized health care.

Looking further into the future, the principles of protein engineering are also underpinning the burgeoning fields of

Synthetic biology involves developing new protein-based materials. Its goal is to design and create new biological parts, devices, and systems or to modify existing natural systems for practical uses. This includes building molecular machines that assemble nanoscale structures, perform complex computations inside cells, or serve as parts of artificial life forms. The ability to precisely control a protein's structure and function through amino acid sequence design greatly expands the possibilities for creating custom biological parts for various applications. For instance, engineered proteins can be designed to self-assemble into detailed three-dimensional scaffolds for tissue engineering, providing a supportive matrix for cell growth and tissue regeneration. These scaffolds can imitate the extracellular matrix of specific organs, guiding cell differentiation and organization.

The journey from understanding the proteome to engineering bespoke proteins is not without its challenges. Accurately predicting protein structure and function from sequence remains a complex problem, although advances in artificial intelligence and computational modeling are rapidly accelerating progress in this area.

Proteins, once seen as mysterious molecular machines with gradually uncovering roles, are now increasingly viewed as programmable entities that can be intentionally designed and used. The combination of proteomics and protein engineering is driving a major change in medicine, moving us toward an era where treatments are not only effective but also precisely tailored to each person. This is not just a small improvement; it marks a fundamental shift in how we approach health and disease. The influence goes well beyond medicine, promising new materials, advanced biotechnologies, and deeper insights into life itself. As our ability to read, understand, and manipulate the language of proteins keeps improving, the future of human health and innovation will undoubtedly be strongly shaped by these remarkable molecular marvels. The endless potential of proteins—ranging from acting as precise tools in personalized medicine to forming the backbone of future biotechnologies—ensures their ongoing central role in uncovering life's mysteries and enhancing human well-being for generations to come.

What started as a simple idea that proteins are essential nutrients and building blocks for muscle and daily health has grown into a study of molecules with incredible complexity. Beyond the dinner plate, proteins are active, adaptable macromolecules whose structures determine their functions, regulation, and even life itself. This hidden complexity, woven into every fold and interaction, is what this book aims to uncover. Ultimately, understanding proteins offers a window into how life organizes matter into meaning and how even the smallest structures reflect the deepest order of the natural world.

Appendix

Allosteric regulation: The control of a protein's function through shape changes caused by molecules binding at a separate regulatory site.

Amino Acid: The building blocks of proteins, characterized by a central carbon atom, an amino group, a carboxyl group, and a variable side chain.

Antibody-Drug Conjugate (ADC): A targeted cancer therapy that links a cytotoxic drug to an antibody that can specifically target cancer cells.

Antibody: A Y-shaped protein produced by the immune system to neutralize pathogens by binding to specific antigens.

Aquaporins: Membrane proteins that form channels allowing water molecules to move rapidly and selectively across cell membranes while preventing the passage of most other substances.

ATP: Adenosine triphosphate. Is the primary energy-carrying molecule in cells. It stores chemical energy in its three phosphate groups and releases that energy when one phosphate group is removed, powering processes such as muscle contraction, active transport, and the synthesis of biomolecules.

Biosensor: A device that combines a biological component (like an engineered protein) with a physicochemical detector to identify specific substances.

Biomarker: A measurable indicator of some biological state or condition, such as the presence of a disease or the response to a treatment.

Breathing motions: Breathing motions are the small, continuous shape changes that allow proteins to function effectively.

Complement System: The complement system is a team of blood proteins that helps the immune system recognize, attack, and eliminate harmful microbes.

Cooperativity: Cooperativity is when one binding event affects the likelihood of future binding events on the same protein.

Directed Evolution: A method used in protein engineering to artificially evolve proteins or nucleic acids to generate molecules with novel or enhanced functions.

ELISA: Stands for Enzyme-Linked Immunosorbent Assay. It is a laboratory test used to detect and measure specific proteins, antibodies, or antigens in a sample, such as blood or other body fluids. It functions by using antibodies and an enzyme that produces a color change when the target substance is present. The more intense the color, the higher the amount of the target substance in the sample.

Entropy: A measure of how much disorder or randomness exists in a system. Left alone, most things naturally move toward greater disorder.

Enzyme: A biological catalyst, usually a protein, that speeds up specific biochemical reactions.

Eukaryotic: Describes cells or organisms whose cells contain a membrane-bound nucleus and other membrane-bound organelles, such as mitochondria and the endoplasmic reticulum. Animals, plants, fungi, and protists are all eukaryotic organisms.

Genome: The complete set of genetic information in an organism.

Isotypes: Isotypes are different forms of the same protein that perform similar jobs but have slight structural differences.

Mass Spectrometry (MS): An analytical technique used to identify and quantify molecules in a sample by measuring the mass-to-charge ratio of ions.

Open reading frame (ORF): A continuous stretch of DNA or RNA that begins with a start codon, ends with a stop codon, and has the potential to be translated into a protein.

Opsonization: The process of tagging a pathogen so immune cells can find and destroy it more efficiently.

Personalized Medicine: A medical approach that tailors disease prevention and treatment to the individual characteristics of each patient, often based on their genetic makeup or proteomic profile.

Prion: An unusual kind of infectious agent made entirely of protein, not DNA or RNA.

Prokaryotic: Cells or organisms that lack a membrane-bound nucleus and other membrane-bound organelles. Their DNA is located in the cytoplasm, and they are generally smaller and simpler than eukaryotic cells. Bacteria and archaea are prokaryotic organisms.

Proteome: The entire set of proteins expressed by a genome, cell, tissue, or organism at a particular time.

Proteomics: The large-scale study of proteomes, including their structure, function, interactions, and quantitative analysis.

Rational Design: A protein engineering approach where modifications are made to a protein's amino acid sequence based on existing knowledge of its structure and function to achieve a desired outcome.

Replisome: A multiprotein complex that replicates DNA at the replication fork. It coordinates DNA unwinding and the synthesis of new DNA strands, ensuring efficient and accurate genome duplication.

Ribosome: A ribosome is a tiny structure in cells that makes proteins by linking amino acids together according to instructions from mRNA.

Synthetic Biology: The design and construction of new biological parts, devices, and systems, or the redesign of existing natural biological systems for useful purposes.

Therapeutic Agent: A drug or substance used to treat a disease or condition.

Zymogens: Also called proenzymes, are inactive precursor forms of enzymes that must undergo a specific chemical change, usually the removal of part of the protein, to become active. This mechanism prevents enzymes from acting before they reach the correct location or at the correct time.

Further Reading

Bai, Yushi, Quan Luo, and Junqiu Liu. "Protein Self-Assembly via Supramolecular Strategies." *Chemical Society Reviews* 45, no. 10 (April 15, 2016): 2756–2767.

Cech, Thomas R. *The Catalyst: "RNA and the Quest to Unlock Life's Deepest Secrets.*" London: W. W. Norton & Company Ltd., 2024.

Chiti, Fabrizio, and Christopher M. Dobson. 2006. "Protein Misfolding, Functional Amyloid, and Human Disease." *Annual Review of Biochemistry*. 75:333-366.

Fallon, Timothy R., Vikram V. Shende, Igor H. Wierzbicki, Amanda L. Pendleton, Nathan F. Watervoort, Robert P. Auber, David J. Gonzalez, Jennifer H. Wisecaver, and Bradley S. Moore. "Giant Polyketide Synthase Enzymes Biosynthesize a Giant Marine Polyether Biotoxin." *Science* 385, no. 6709 (August 9, 2024): 671–678.

Guziewicz NA, Massetti AJ, Perez-Ramirez BJ, Kaplan DL. "Mechanisms of monoclonal antibody stabilization and release from silk biomaterials". *Biomaterials.* 2013 Oct;34(31):7766-75. doi: 10.1016/j.biomaterials.2013.06.039. Epub 2013 Jul 13. PMID: 23859659; PMCID: PMC3756937.

Holyoak, T. (2013). "Molecular Recognition: Lock-and-Key, Induced Fit, and Conformational Selection". In: Roberts, G.C.K. (eds) *Encyclopedia of Biophysics*. Springer, Berlin, Heidelberg. https://doi.org/10.1007/978-3-642-16712-6_468

Koshland, D.E., Jr. (1995), "The Key–Lock Theory and the Induced Fit Theory". *Angew. Chem. Int.* Ed. Engl., 33: 2375-2378.

Levinthal, C. (1968). "Are there Pathways for protein folding?" *Journal of Chemical Physics*, 65, 44-45

Matsuura, Kazunori. "Rational Design of Self-Assembled Proteins and Peptides for Nano- and Micro-Sized Architectures." *RSC Advances* 4 (2014): 2942–2953.

Morin, Randy, David Kaplan, and Bernardo Perez-Ramirez. "Bone Morphogenetic Protein-2 Binds as Multilayers to a Collagen Delivery Matrix: An Equilibrium Thermodynamic Analysis." *Biomacromolecules* 7, no. 1 (2006): 131–138.

Pauling, L., Corey, R. B., & Branson, H. R. (1951). "The structure of proteins: two hydrogen-bonded helical configurations of the polypeptide chain." *Proceedings of the National Academy of Sciences of the United States of America,* 37(4), 205–211.

Pérez-Ramírez, Bernardo. *The Engines of Life: Structure and Function of Proteins at Work.* Tampa, FL: Gatekeeper Press, 2024.

Perez-Ramirez, Bernardo, Keith E. Shearwin, and Serge N. Timasheff. 1994. "The Colchicine-Induced GTPase Activity of

Tubulin: State of the Product. Activation by Microtubule-Promoting Cosolvents." *Biochemistry* 33 (20): 6253–61.

Perez-Ramirez B. "Thermal stability of Torpedo californica acetylcholine receptor in a cholesterol lipid environment." *Mol Cell Biochem.* 1994 Mar 30;132(2):91-9.

Rahbar S. "The discovery of glycated hemoglobin: a major event in the study of nonenzymatic chemistry in biological systems." *Ann N Y Acad Sci.* 2005 Jun;1043:9-19. doi: 10.1196/annals.1333.002. PMID: 16037217.

Senior AE, Nadanaciva S, Weber J. "The molecular mechanism of ATP synthesis by F1F0-ATP synthase." *Biochim Biophys Acta.* 2002 Feb 15;1553(3):188-211.

Schechter AN. "Hemoglobin research and the origins of molecular medicine." *Blood.* 2008 Nov 15;112(10):3927-38. doi: 10.1182/blood-2008-04-078188. PMID: 18988877; PMCID: PMC2581994.

Snowdon, David. 2001. *Aging with Grace: What the Nun Study Teaches Us About Leading Longer, Healthier, and More Meaningful Lives.* New York: Bantam Books.

Vera, J. C., M. Brito, T. Zuvic, and L. O. Burzio. "Polypeptide Composition of Rat Sperm Outer Dense Fibers: A Simple Procedure to Isolate the Fibrillar Complex." *Journal of Biological Chemistry* 259, no. 9 (1984): 5970–5977.

Westring DW, Grand S. "Screening for sickle hemoglobin--a review." *Am J Med Sci.* 1973 May;265(5):358-66. doi: 10.1097/00000441-197305000-00001. PMID: 4576755.

Young, E. C., E. Berliner, H. K. Mahtani, B. Perez-Ramirez, and J. Gelles. 1995. "Subunit Interactions in Dimeric Kinesin Heavy Chain Derivatives That Lack the Kinesin Rod." *Journal of Biological Chemistry* 270 (8): 3926–3931. https://doi.org/10.1074/jbc.270.8.3926.

Zhu, Jie, Nicole Avakyan, Albert Kakkis, Alexander M. Hoffnagle, Kenneth Han, Yiying Li, Zhiyin Zhang, Tae Su Choi, Youjeong Na, Chung-Jui Yu, and F. Akif Tezcan. "Protein Assembly by Design." *Chemical Reviews* 121, no. 22 (August 18, 2021): 13701–13796.

About the Author

Bernardo Pérez de Buerres Ramírez is a scientist, biopharmaceutical executive, professor, writer, and philanthropist. He holds a PhD in biochemistry and biophysics of proteins and has over thirty years of experience in the structure and function of proteins in solution, as well as in the development of formulations for therapeutic proteins. He is an adjunct professor in the Department of Biomedical Engineering at Tufts University in Boston and an honorary professor at the University of Nottingham in the United Kingdom.

Dr. Pérez de Buerres Ramírez is the author of more than 40 scientific articles and several books on topics including protein science, history, heraldry, poetry, novels, and essays. He is a Fellow of the Royal Society of Chemistry of the United Kingdom, an elected member of the American Society of Biochemistry and Molecular Biology, a member of the Spanish Society of Biochemistry and Molecular Biology, and an Academician of the Instituto de Estudios Históricos Bancés y Valdéz- Spain.

www.ingramcontent.com/pod-product-compliance
Lightning Source LLC
LaVergne TN
LVHW090401210926
848480LV00002B/411
9798993187228